Organization Management in Construction

T0265060

Today's construction environment is changing at an unprecedented pace and executives are facing a diverse set of issues – globalization, economics, workforce demographics, and technology. Moreover, the traditional issues of competition and delivery are being challenged by new laws and new industry entrants, and the tasks of project and organization management are being overhauled. This all demands greater leadership from senior management.

Construction executives typically reach senior level after many years mastering the art of project management, which has given them very little time or opportunity to learn the concepts and principles of organization leadership – unlike their counterparts in other industries who have been steeped in this.

This book provides a comprehensive overview of the key issues that organization leaders must understand and address. It provides concise summaries by leading international authorities of the ten key strategic management issues, shows how they have emerged, and outlines their potential impact on the construction organization.

Paul S. Chinowsky is Professor at the University of Colorado, USA.

Anthony D. Songer is Professor at Boise State University, Idaho, USA.

Organization Management in Construction

Edited by Paul S. Chinowsky and
Anthony D. Songer

Routledge
Taylor & Francis Group

LONDON AND NEW YORK

First published 2011
by Spon Press

Published 2015
by Routledge
2 Park Square, Milton Park, Abingdon, Oxon OX14 4RN

Simultaneously published in the USA and Canada
by Routledge
711 Third Avenue, New York, NY 10017

Routledge is an imprint of the Taylor & Francis Group, an informa business

Typeset in Goudy by
Saxon Graphics Ltd, Derby

British Library Cataloguing in Publication Data
A catalogue record for this book is available from the British Library

Library of Congress Cataloging-in-Publication Data
Organization management in construction / edited by Paul Chinowsky and Anthony Songer.
   p. cm.
   1. Construction industry--Management. I. Chinowsky, Paul. II. Songer, Anthony.
   HD9715.A2O72 2011
   624.068--dc22
   2010032995

ISBN13: 978-0-415-57257-6 (hbk)
ISBN13: 978-0-415-57261-3 (pbk)
ISBN13: 978-0-203-85610-9 (ebk)

Contents

Illustrations

Figures

Tables

Contributors

C. Anac, University of Reading, UK

M.T. Birgonul, Middle East Technical University, Ankara, Turkey

Paul Chan, University of Manchester, UK

Paul S. Chinowsky, University of Colorado, USA

Andy Dainty, Loughborough University, UK

Steve Darr, Director, Peacework Development Fund, Inc., Virginia, USA

I. Dikmen, Middle East Technical University, Ankara, Turkey

Michael J. Garvin, Virginia Tech, Virginia, USA

Matthew R. Hallowell, University of Colorado, USA

Timo Hartmann, Twente University, The Netherlands

Jimmie Hinze, University of Florida, USA

Amy Javernick-Will, University of Colorado, Boulder, CO, USA

Ashwin Mahalingham, IIT Madras, India

S. Ping, National Taiwan University, Taiwan

Anthony D. Songer, Boise State University, Idaho, USA

T. Michael Toole, Bucknell University, Pennsylvania, USA

Preface

It seems like the only constant in the world in which we live is that change comes faster each year. Each year computers become faster, lighter, smaller, and smarter. The concept of a phone is changing so rapidly that calling today's cellular device a phone is almost demeaning to the manufacturer. The concept of friends is even changing as a friend may mean an individual who you have lunch with on a regular basis to somebody you have never met in person, but with whom you chat on a regular basis. Even the concepts of work, play, and work-life balance are changing as communications now allow an individual to remain plugged in whether they are in the office, on a golf course, at the beach, or even on a chairlift. We live in a 24 hour world that never turns off, never slows down, and follows us wherever we go.

This is an environment that many construction leaders were not trained for, did not begin their careers in, and did not imagine they would be leading a firm in. However, this is the world in which we live and just as in every age for the last 3,000 years, it is not looking back. Therefore, we are forced to confront the issue of what are the relevant topics to construction professionals today? What are the key issues that organizations are going to face? How does a current or future leader ensure that their organization is going to be sustainable and successful over the long-term?

These are the questions that drove the writing of this book. At the beginning of any new era, one needs to look at how the new era is going to affect the current organization. Similar to the beginning of every previous era, we are looking at a time when we think the coming changes are overwhelming and present a never before seen set of challenges. On the positive side, we can say that although there are in fact significant changes emerging, they are not in any way more challenging than issues of the past. Think of the introduction of computers and how that indeed changed the industry, but the industry survived. However, on the negative side, the current set of changes requires industry leaders to make changes to their organizations or risk being swept aside in a tide of globalization.

This book challenges both current and future leaders to examine these issues and make changes to accommodate the changes ahead of the wave to avoid being swept under by the wave.

The chapters

Within each chapter of the book, the authors present the key issues that are facing organization leaders, where they originated and the potential impact on the construction organization. Organization leaders and individuals who are preparing for a leadership position will gain the breadth of understanding required to successfully engage in developing solutions to the challenges facing the construction organization of today.

The book falls broadly into two sections, internal and external issues, each of which contains chapters that build a perspective for enhancing organization success from the foundation to the application. Beginning with the internal issues, the first topic is the overall need for leadership as the foundation for organization success. Tony Songer integrates leadership theory with the need to rethink the roles of a leader in the changing economic environment. Building on the new leadership requirements, Amy Javernick-Will and Timo Hartmann address the need to rethink the manner in which organizations share and distribute knowledge. Focusing on knowledge as a key element for long-term success, Javernick-Will and Hartmann explore the challenges and barriers with knowledge management as well as the opportunities for organizations that emphasize this strategic area. Building such a knowledge network requires individuals to work as a team in an effort to translate knowledge to performance. This transformation process is the focus of Paul Chinowsky's chapter on networks as the key to achieving high performance.

Continuing with the internal issues, Michael Toole, Paul Chinowsky, and Matthew Hallowell introduce the need for organizations to increase the focus on innovation as a strategic advantage. Based on research that demonstrates the long-tem advantage of investment in innovation, the chapter provides a guideline for embarking on such a strategic transformation. Completing the focus on internal issues is the chapter by Andrew Dainty and Paul Chan on human resource issues. The chapter is foundational for all organizations as it makes it clear that any internal or external endeavor is limited by the degree to which personnel in an organization are committed to both their individual and collective success.

In the second half of the book, the emphasis changes to external issues. The first of these issues is the emergence of global markets. S. Ping Ho emphasizes the rise of the global market as the next challenge and opportunity for construction organizations as they reorganize to address the new realities of the marketplace. Following this same path, Irem Dikmen and Talat Birgonul address the emerging markets that are changing the traditional concepts of construction services. The combination of these two chapters provides a starting point for any organization to commence the discussion of where the organization is headed both near-term and long-term.

Moving away from the emerging market focus to one that is more service focused, Michael Garvin and Ashwin Mahalingam introduce the new finance methods that are emerging on a global basis. Once considered alternative methods that are employed on a limited basis, finance alternatives are rapidly becoming standard practices with rapid introduction in the developing world. Continuing with this focus on new project thinking, Matthew Hallowell and Jimmie Hinze introduce the importance of safety as a strategic asset to an organization. With an increased emphasis on global projects, this topic is moving away from the exclusive domain of project managers to a strategic topic that impacts overall strategic success.

Finally, the focus on emerging issues concludes with Steve Darr and Tony Songer addressing the role of corporate social responsibility. In an age where global engagement is the focus of the emerging markets, understanding the role that organizations can and should play in enhancing the developing world is a key concern for organization leaders. This chapter addresses the keys to initiating such endeavors and establishing the strategic concerns for successfully becoming a global citizen.

Finally, Paul Chinowsky and Tony Songer bring the book to a close by synthesizing the lessons introduced in the book and asking the key questions that executives need to address as they embark on a path to succeeding in the rapidly changing economic environment.

To the reader

This book can be read from different perspectives. One perspective is that of the student examining the new influences on the construction industry. To those readers we invite you to take a critical look at the industry around you and challenge leaders on how they are addressing the points raised in this volume. A second perspective is that of the aspiring organization leader trying to understand the new challenges that await his or her ascension to senior leadership authority. To that reader we invite you to use this book as a stepping stone to obtain further knowledge on the topics that you see as critical to your organization. This book is the tip of the iceberg. Gather more knowledge to avoid a shipwreck in your future. Finally, there is the perspective of the current leader looking for a starting point to a strategic discussion. To these readers, we encourage you to use this book as a starting point for strategic discussions with the leadership team in terms of how you are addressing the new economic environment. Your attention to these issues may be the determinant of your long-term success or a far worse alternative.

Paul S. Chinowsky
Anthony D. Songer

1 Introduction

Paul S. Chinowsky and Anthony D. Songer

The engineering-construction industry is one of the driving forces of the world economy with $3.2 trillion in expenditures worldwide (ENR 2000). However, it is an industry with business management practices that need improving, exemplified by a business failure rate that is 34 percent higher than the national average for all industries (Statistical Abstract of the United States 2000). It is an industry organized by size with an expansive gap between the upper echelon of performers and the smaller firms. The industry is characterized by changing forces in technology, economics, marketing, politics, and its workforce. Furthermore, it is an industry plagued by a persistent negative image, despite its economic force.

Concurrent to this negative image, the construction industry continues to stratify as illustrated by the revenue distribution between the small and large contactors. For example in the United States, there are approximately 199,286 construction organizations (US Bureau of the Census 2000). However, 46 percent of the revenue is being accumulated by less than 1 percent of the industry. Adding to this stratification is the increase in global competition affecting engineering and construction organizations in every region of the globe. In today's marketplace, international competition is increasing as technology is changing the face of industry competition. Specifically, technology is reducing the importance of physical location.

With this increase in competition, engineering and construction, leaders must place additional emphasis on long-term success and survival. However, the fundamental component of the industry, the workforce, is also changing. Technology, immigration, education, and the standard of living are changing the dynamics of this workforce. Furthermore, the construction industry is seen as tedious, dirty, non-technical, non-professional, hazardous, cyclical, and associated with difficult working conditions (Abraham 2002). Despite the numerous construction achievements throughout the world, the construction industry receives one of the lowest ratings for jobs. The 1999 Jobs Rated Almanac ranked the job of a construction worker as 247 out of 250, followed by a fisherman, a lumberjack, and a roustabout (Bodapati and Naney 2001). Compounding this image problem is a low productivity rating. With a productivity growth rate of only 0.8 percent from 1970 to 1986, the construction industry ranked at the bottom of all industries examined during that period (Adrian 1987). Additionally, the construction industry is one of the most dangerous industries. Construction has a disproportionately high level of deaths and injuries, accounting for over 21 percent of deaths in industrial accidents in the United States, while only employing 8 percent of the workforce (CPWR 2007). Finally, after contributing approximately 8 percent of the gross domestic product in the United States, only 0.5 percent of that was expended for construction research, leading to a perception that the industry lacks the ambition to improve and innovate (Bodapati and Naney 2001). While these statistics are from the United States, the global perspective on the industry does not differ significantly.

These changes and forces necessitate altering the traditional perspective of the industry, both from professionals within and observers outside the industry. No longer can industry enterprises be satisfied with continuing long-held family or corporate traditions. Similarly, universities can no longer view the education of future professionals in the same manner. Rather, the question for both industry and academia is how to address the engineering profession and the leaders of the future. What tools are required to successfully lead a construction industry organization? What should be the primary focus of organization leaders? Will these individuals be the integrators of the design-construction process, or will they be the technology experts focusing on satisfying customer requirements, or maybe they will be the surrogate owners, financing, building, and operating projects on an international scale.

What these changes are paving the way for, and what these questions are highlighting is the need for industry leaders to re-examine their individual roles and the influences on the industry. It is becoming increasingly clear that industry leaders will need to adopt a more comprehensive approach to leading an organization. Specifically, leaders will need to move toward an integrator concept where an industry leader is a life-cycle integrator. In this concept, the leader emphasizes the ability to integrate the objectives, plans, and requirements of the business, design, owner, and construction constituencies into a single coherent effort. However, to achieve this integrator profile, the professional will be required to have balance in their personal education, professional development, and knowledge of available resources. Organization leaders will need to balance formal education received in school and through professional development with practical intelligence received through professional experience, and emotional intelligence associated with psychological, sociological, and interpersonal communications and relationships. Currently, this profile is often in a state of imbalance as the engineering leader accentuates a single area of the spectrum of knowledge required to fully charge the knowledge bases that support these three profile components.

In summary, the next generation of industry leader will need the breadth of knowledge to address diverse concerns throughout the project life cycle. The executive must be problem solver, leader, manager of change and relationships. Additionally, this professional will need the balanced profile required to succeed in the new global economy. And finally, this professional will need the tools and education required to proactively address the continued need for advancement within the construction profession.

Current state issues

Changing the profile of an industry leader is not a simple task. If anything, the construction industry is one based in tradition. Change must be associated with a compelling reason and/or a significant consequence. Although we are not striving to provide this compelling reason directly, we do present in this book the drivers that are establishing the compelling need for change. As an introduction, the issues driving the change in the current direction of the construction engineering and management domain can be summarized as follows.

- **Integration vs. specialization** – In a strong move away from the trend witnessed over the last half century, the next generation of industry leaders will need to balance the increasingly disparate and specialized views expressed within the context of a project team. Specifically, views from architects, financiers, risk specialists, engineers, owners and others are often in conflict throughout the project timeframe. These conflicts need to be balanced and resolved to ensure the completion of the project that is every participant's ultimate goal. However, the manner in which this goal is achieved can take many paths including confrontational approaches, coordinated approaches, and balanced approaches, among others. The question for the future of the construction industry is how to reduce the focus on confrontation and increase the focus on collaboration and balance. Rather than focus on increasing knowledge specialization, leaders will need a perspective and context to address the potentially divergent goals represented on a project and weave these perspectives into a unified set of goals and directions.

- **Successful intelligence model** – The predominant perspective on construction preparation for success has centered on the mastery of practical knowledge. In this perspective, construction professionals were rewarded for their extensive personal knowledge base of operations and project management. The greater that an individual could display experience and ability to plan and organize field operations, the greater the opportunity for long-term advancement within a construction organization. This perspective of construction knowledge development has been slowly changing as a broader success model is emerging that reflects the need for integration. In this context, construction professionals will develop knowledge in each of the successful intelligence areas; practical, emotional, and intellectual intelligence.

- **Body of knowledge vs. individualism** – What do construction professionals do? What is the definition of a construction professional? What will the construction industry look like in 2050? What services will the construction professional provide in 2050? These questions center on a critical issue for the construction industry – what is the body of knowledge for a construction professional? The lack of consensus on this issue has significant repercussions for every aspect of the construction industry. For the academic side, the lack of an agreed-upon body of knowledge makes it impossible for academic programs to move toward an agreed-upon curriculum. For construction professionals, the lack of this body of knowledge makes it difficult to define what services should be offered by the construction organization. And, for the owner, the lack of this body of knowledge consensus makes it difficult to set expectations for services on individual projects. Although different associations are attempting to address this issue, the changing global environment is making the decision of what is essential knowledge a question that needs to be addressed by each organization.

- **Industry-academia divide** – Industry and academia are dependent on each other for long-term success. Closer collaboration is required for future industry development. However, industry-academic collaboration has become an exception rather than an expected norm. The reasons for this divide are varied, including lack of trust, lack of respect for respective experience and knowledge, and lack of recognition for respective roles. The result of this divide is clear. Industry professionals and academics are losing the opportunity to enhance their respective professional roles through collaboration. Additionally, the move toward a new generation of construction professional requires each side of the divide to recognize the importance of contributions made by the construction community as a whole.

Although addressing these issues does not guarantee the success of an organization, it does represent a first-step towards an acknowledgement that the construction industry is changing and with it will change the concept of the construction professional. However, the authors do not want to leave the reader with the perception that simply by moving down the road toward addressing these issues will be a final solution. Rather, issues such as measurement of success come into play once an organization moves forward with internal and external changes. Specifically, an organization must evaluate the success of this new perspective using tangible measurements. The substance of these measurements should be a critical priority for the organization as it looks to the future.

Organization impact

What does this change mean to industry organizations? Do organizations need to abandon their core competencies and learn a new area of services? The direct answer is no. However, some alterations will be required. The comfort of knowing who your competition is likely to be before proposals are even submitted is likely to diminish. The opportunity to focus extensively on a narrow band of services is going to diminish. The traditional perspective of growing personnel from entry level positions through a 25 year career to executive positions is already fading quickly. All of these things are going to force change in the traditional construction industry environment. Compounding this issue is the change in the owner perspective. Put simply, owners are getting smarter, owners are getting more demanding, and owners are looking for partners who understand their business and understand how you can make a difference for them. Taken together, this will alter the future look of the engineering or construction office.

At the core of this change is going to be the way in which leaders guide the organization. The new organization is going to require leaders who are more hands-on in their approach while also bringing a greater level of trust to the organization. It is going to take leaders who understand that their value is in the ability to synthesize the "big picture". Spending time developing new forms to track corporate credit card spending is not where a leader should be spending his or her time. Similarly, understanding that breadth of knowledge is as important as depth of knowledge within the firm is critically important to achieve the balance discussed earlier. Every individual in the organization does not have to have the same background or set of skills. Every person does not have to go through the same training course. Diversity and balance are going to be the keywords of the new corporate environment.

Do all these things add up to changes in the overall organizations? Yes, there is going to be change required and those who adapt to this change will have a sustainable advantage over the competition. Fundamental issues such as education plans, career paths, and leadership selection will need to be re-examined in terms of how they are developed and executed. Strategic thinking will need to make a comeback in terms of setting long-term goals and determining how the organization should be perceived from the outside. The era of living comfortably on existing clients and limited competition is going to close. The organizations that are able to adapt to this are the ones that will succeed in the new environment.

As a final consideration, these changes cannot be completely controlled by organization leaders. A successful transition will require the commitment and work of all individuals in the organization. Without opening up a controversy in terms of education and what is appropriate for employees to learn before and during their careers, we will say that organizations need to examine what type of employees they are hiring, what are their educational backgrounds, what do they want to achieve, and what is their understanding of the types of clients the organization services. It is always important to remember that the weakest link in an organization is the employee who is speaking with a client and does not know the overall goals and direction of the organization.

How do we proceed?

Organization structure is changing. Leadership requirements are changing. The environment in which construction organizations operate is changing. With all of this change, how should organizations in the construction industry adapt and proceed? Unfortunately, the response to this situation is not a single entity solution. A part of the solution is adaptations that are required in the education system. It may not be appropriate in the long term to have students focus on a narrow area within the industry. It may be appropriate to introduce new curricula in international affairs, leadership, business, and management, among other items. However, this aspect of the discussion is outside the scope of this book. What is in the scope of this book is the adaptation from the organization perspective to the new challenges facing the industry. In this context, we must focus on the two remaining, distinct components of the solution; internal organization elements and external organization elements.

Internal organization elements are those elements that can be addressed by organization leaders in terms of how the organization is going to operate and position itself against the competition. Elements in this category include the strategies that the organization is going to put in place, the leadership that is going to be employed, the education that is going to be provided, and the financial decisions that are going to be made.

External organization elements are those elements that need to be addressed in the context of the environment in which the organization is operating. For example, the competitive environment, the global market opportunities, the evolving safety demands, and the need for greater knowledge sharing in interorganizational teams. All of these issues require an understanding of and a focus on, the external environment.

Building the understanding of how to address the key issues in each of these areas is the focus of this book. Experts in construction organization issues from around the globe were brought together to bring their insights, experience, and research results into a single volume. Is every issue that emerges in the internal and external areas covered in the book? No, that would be difficult if not impossible to achieve given the limited space of any volume. However, the book contains a set of key issues that the authors have agreed upon that are emerging as critical near-term issues for construction executives. The chapters should be read in terms of this context. They are not a comprehensive set, but they represent the next stage in organization management critical issues.

In summary, this book provides a comprehensive introduction to the roles, concerns, and issues that are facing the construction industry executive and the impacts that these issues will have on the industry. Utilizing international authorities on each subject, the book fills a gap in the knowledge base of current and future industry executives.

References

Abraham, G. (2002) *Critical Success Factors for the Construction Industry*, PhD dissertation, Georgia Institute of Technology (advisor: Paul Chinowsky), July 2002.

Adrian, J.J. (1987) *Construction Productivity Improvement*, Upper Saddle River, NJ: Prentice-Hall Publishers.

Bodapati, S.N. and Naney, D. (2001) "A perspective on the image of the construction industry," *Associate Schools of Construction International Proceedings of the 37th Annual Conference*, pp. 213-223.

CPWR (2007) *The Construction Chart Book*, Silver Spring, MD: The Center for Construction Research and Training.

ENR (2000) *The Top 500 Design Firms Sourcebook*, New York: McGraw-Hill.

Statistical Abstract of the United States (2000) U.S. Department of Commerce, Bureau of the Census, Washington, DC.

US Bureau of the Census (2000) *1997 Economic Census*, U.S. Department of Commerce, Bureau of the Census, Washington, DC.

2 Leading the modern construction organization

Anthony D. Songer and Paul S. Chinowsky

Introduction

Construction industry organizations continue to change at a rapid pace. Influences such as globalization, evolving delivery mechanisms, and changing organizational structures require business decisions that challenge the traditional transactional focus of the industry. One increasingly important issue facing the industry is that of leader preparedness. Recent publications by the National Academy of Engineering and the American Society of Civil Engineers document the significance of leadership development to the industry. While certainly not a new phenomenon, there continues to be a shortage of leadership training within the US construction industry. Only 10 percent of companies provide any leadership development (Skipper and Bell 2006; Bogus 2006).

This lack of focus on leadership training is leading to the concern as to where the leaders for the decade are going to emerge. The determination of which individuals possess the knowledge and skills to address rapidly changing business environments is becoming more difficult as the focus on leadership declines. Leadership ability to transform organizations in response to dynamic market forces is essential for sustained success within the construction industry. Additionally, the ability for leaders to motivate the internal workforce and communicate to a diverse project team requires heightened interpersonal communication skills. This chapter addresses this concern as it:

1 establishes the leadership needs and challenges specific to the construction industry,
2 describes the integrator model as a framework for developing the next generation leader skills, and
3 provides a detailed discussion on two leadership topics: emotional competency and collaborative acumen.

Leadership needs

A study by the authors on current construction industry leaders identified both attributes of leaders as well as the primary challenges facing the industry in terms of leadership.

Attributes of leaders

Construction industry leaders believe strongly that the three most important traits of a leader are integrity, the ability to interact with others either through communication or interpersonal relationships, and the ability to set a corporate vision. (Songer and Chinowsky

2008). Integrity in leadership is not specific to the construction industry. Integrity and its link to honesty appears in most discussions on attributes of leaders (Kouzes and Posner 2002). What is interesting is that construction industry leaders consider the perceived lack of integrity among leaders as a major issue for the next generation of leaders.

The second leadership attribute, communication, is also widely documented with organizational leaders consistently recognizing this trait as a key to leadership success. However, communication is becoming increasingly important as construction leaders begin to understand that issues such as increased competition require additional client communication to retain client loyalty as well as the evolution of increasingly collaborative delivery mechanisms. Interestingly, this is occurring at the same time as technology innovations increasingly provide collaboration tools which decrease the necessity for human interaction. Thus, an emphasis on communication is rising to levels beyond the traditional emphasis.

The final leadership attribute, vision, is perhaps the most unexpected of the top three attributes as expressed by construction executives. A leader who is visionary and can set long-term strategic goals is essential for the organization. While literature indentifies this as an essential component of leadership, it was unexpected in that developing vision is given far less emphasis in practice and leadership development than technical responsibilities. Further, the conservative nature of the industry often focuses more on responding to short-term client demands than setting visions. This trait is revisited later as the question arises as to how an organization develops visionary leaders when the industry has a strong bias towards project or transactional decisions.

Industry challenges

Many challenges facing the construction industry exhibit a strong focus on leadership and workforce challenges. The principle challenges of lack of quality personnel, attracting talent, aging workforce, workforce issues, and training, indicate both a need for leadership and labor development. Specifically, the predominance of workforce issues is not isolated to labor, it includes the shortage and need of mid and senior level leadership training.

A particular concern in the industry is the availability of personnel who demonstrate strong leadership characteristics as well as having the ability to motivate personnel both in the office and at the job site. As related to leadership attributes, there is a clear concern that individuals displaying the attributes such as integrity, communication, and vision may not be available to the industry in the numbers required to retain a strong competitive position.

To address the needs and challenges facing the construction industry the authors developed a new model for the future industry leader, the integrator model. The integrator model illustrates the three competency areas required for next generation leadership. After a brief description of the integrator model, this chapter refines the discussion toward the emotionally competent leader and opportunities to lead in the emerging organizations of integrated project delivery.

The integrator model

As noted, current changes in organizations, technologies, and markets necessitate altering the traditional perspective of the construction industry, both from professionals within and observers outside the industry. No longer can industry enterprises be satisfied with continuing long-held family or corporate traditions. Similarly, universities can no longer view the education of future professionals in the same manner. Rather, the question for both industry and academia is how to address the construction engineering profession and professionals of the future. Where will these professionals emerge? How will they be educated? What tools will they use? What will be their primary focus as professionals? Will these individuals be the integrators of the design-construction process, or will they be the technology experts focusing on satisfying customer requirements, or maybe they will be the surrogate owners, financing, building, and operating projects on an international scale?

The leader as integrator

The integrator concept was developed as a top-down model to begin the process of addressing developing an understanding of next generation construction leadership roles and skills. The top level of the model is the concept of the construction leader as a life-cycle integrator. In this concept, the construction leader emphasizes the ability to integrate the objectives, plans, and requirements of the business, design, owner, and construction constituencies into a single coherent effort. The idea in this concept is that the construction leader will be equipped with the perspective, tools, and responsibility that spans the life cycle of the project.

To achieve this integrator profile, the construction leader must have balance in their education, professional development, and knowledge of available resources. The integrator model illustrates three distinct areas of intelligence: intellectual intelligence which represents the formal education received in school and through professional development; practical intelligence which represents the knowledge received through professional experience; and emotional intelligence which represents the knowledge associated with psychological, sociological, and interpersonal communications and relationships. Currently, this profile is in a state of imbalance as the construction professional fails to receive the spectrum of knowledge required to fully charge the knowledge bases that support these three profile components.

Supporting the three areas of intelligence in the integrator model are three distinct knowledge bases, each correlated to a single area. Within these knowledge bases are "focus areas" that represent the specific topics that the construction professional is required to understand as a component of the knowledge base. The diversity of the focus areas in the cumulative profile illustrates the breadth of knowledge required to fulfill the life-cycle integrator concept. Figure 2.1 illustrates the top level of the integrator model. A detailed description of the model can be found in Songer *et al.* (2002).

This leader as integrator will have the breadth of knowledge to address diverse concerns throughout the project life cycle. The integrator must be problem solver, leader, manager of change and relationships. Additionally, this leader will have the balanced profile required to succeed in the new global economy. And finally, this leader will have the tools and education required to proactively address the continued need for advancement within the construction profession. Hence, the next generation construction leader will:

- perform a fundamentally different task from that performed by a general contractor or construction manager today. The changes in the construction industry are forcing a transition toward a central player that has the capacity to interact with all of the constituents on the project.
- implement a move toward a new definition of services for an owner that may resemble a combination of construction management, general contractor, and construction consulting services that are performed today.

- be a generalist, or "integrator", who has the experience to interact with all project participants, but understands the need for specialists to address specific issues within a project.
- have a greater focus on proactive problem solving, understanding potential conflicts and setting paths that will minimize long-term conflicts.
- require faster immersion into all aspects of an organization than today's entry-level position. It will no longer be feasible to require an individual to track requests for information (RFIs) for 18 months when they begin. Rather, an integrated training program that introduces an individual to every aspect of the organization will be required.
- be a problem solver, team builder, and leader. Focus must be directed on the development of leaders rather than skill sets.

Relevance to organization development

The development of the integrator model provides a foundation on which to examine the potential impact of the model on organization development. First, in the area of new employee development, the critical impact of the integrator concept is that construction education programs do not meet the impending future needs of the industry. With the engineering and management education process rooted in its 1960s origins, the focus of programs is on enhancing specialized knowledge and skills for the construction workforce. However, this paradigm does not match the integrator requirements. Specifically, the current education focus is too narrow. The role of the integrator will require an individual to examine issues from both a life-cycle perspective and from multiple perspectives. Today's curriculum places too much emphasis on the construction planning and execution phases to develop these external concerns. Industry organizations will need to consider this disconnect when deciding on the appropriate professionals to hire for the firm.

In addition to these focal points for individual learning, the integrator concept impacts lifelong learning ideals. The move to a role of integrator will require construction professionals to continue their education beyond the realm of the university. The diversity and breadth of issues associated with integration will require the industry professional to actively pursue education throughout his or her career. To support this requirement, companies must alter their perspective on education to acknowledge the importance of education throughout a professional's career. Although this perspective exists sporadically in the construction industry, it is far from prevalent.

Finally, the less-technical aspects of education must be given greater emphasis in the organization. The integrator model outlines a need for emotional intelligence and organization issues to be incorporated into the profile of future leader success. However, this focus is absent in many areas of today's professional education process. Very few industry organizations are actively incorporating these topics in current leadership programs. However, this focus must be enhanced and recognized as a critical component of the education process. The less-technical aspects of construction and engineering must be given greater emphasis. The integrator model clearly outlines a need for emotional intelligence and organization issues to be incorporated into current leadership development models. The subsequent sections of this chapter focus on the emotional knowledge component of the integrator model, specifically the emotionally competent and collaborative leader competencies that have been grossly overlooked in traditional leader development models.

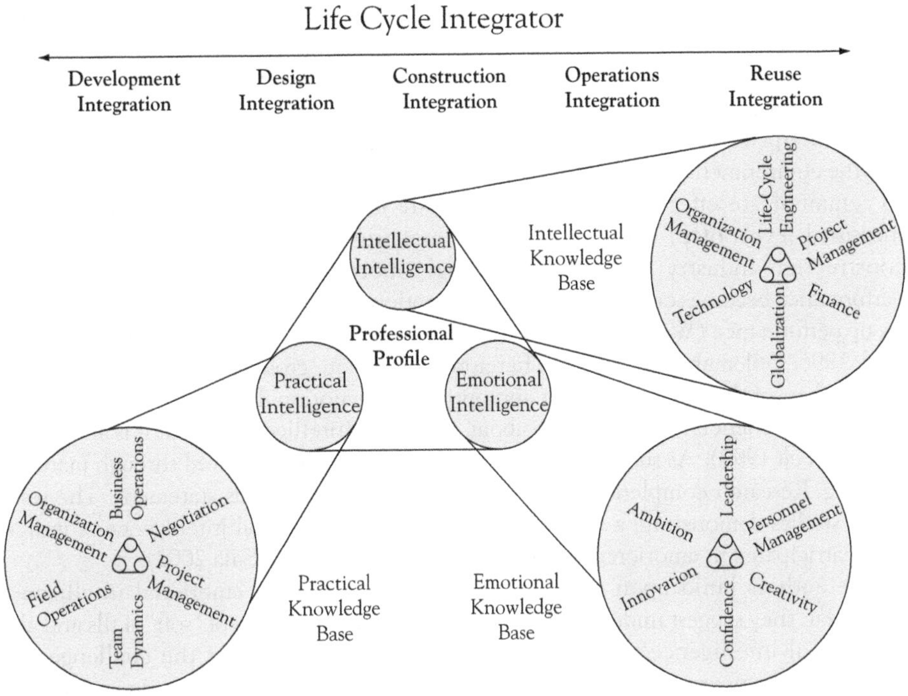

Life Cycle Integrator

| Development Integration | Design Integration | Construction Integration | Operations Integration | Reuse Integration |

Figure 2.1 The integrator model with knowledge bases, intelligences, and the life-cycle concept.

The emotionally competent leader

Emotional intelligence refers to an individual's ability to identify emotions in oneself and others and to exhibit appropriate responses to environmental stimuli. Investigating emotional intelligence among construction executives establishes a foundation for understanding the underlying obstacles of effective leadership.

In recent years, emotional intelligence has gained considerable popular attention. Much of this can be attributed to the popular book *Working with Emotional Intelligence* written by Daniel Goleman (1998). In his book, Goleman makes several strong claims about the contributions of emotional intelligence to the individual and society. As a result of these claims, the general conception of emotional intelligence has become commonly known and has appeared in many magazines and newspaper articles (Mayer *et al.* 2000).

Recent research indicates that the AEC (architecture-engineering-construction) industry is beginning to realize the critical impact that "soft" skills have on the success of projects (Thamhain 1992; Pocock *et al.* 1997; Johnson and Singh 1998; Loosemore 1998; Thomas *et al.* 1998; Black *et al.* 2000; Bresnen and Marshall 2000a, 2000b; Cheng and Li 2001; Carr *et al.* 2002; Ling 2002; Singh 2002). Much of the research conducted on emotional intelligence focuses on how it relates to workplace success. More specifically, many studies analyze the impact of emotional intelligence on individuals' leadership ability and job performance (Scheusner 2002).

Leadership greatly impacts many aspects of organizational effectiveness, which includes production, quality, efficiency, flexibility, satisfaction, competitiveness, and development (Gibson *et al.* 2003). This makes it important to search for what distinguishes outstanding leaders. Emotional intelligence accounts for close to 90 percent of what sets apart exceptional leaders from those deemed as average (Kemper 1999). Additionally, in a study involving 49 top business leaders, three of the top five skills identified as key leadership skills relate to the interpersonal components of emotional intelligence (Scheusner 2002).

The interpersonal aspect of emotional intelligence focuses on how people relate to others and the environment. To ensure top performance of an organization, employees must be able to communicate effectively. Organizations are now demanding that employees possess increased levels of EI skills as teamwork and trustworthiness (Kemper 1999). As the construction industry becomes increasingly more collaborative, teamwork or group performance becomes critical. Fortunately, emotional intelligence has been linked to higher group performance (Williams and Sternberg 1988; Ashforth and Humphrey 1995; Campion et al. 1996; Goleman et al. 2001). Therefore, to promote group effectiveness it is important for members of the group to exhibit appropriate and emotionally intelligent responses.

Another beneficial aspect to note about emotional intelligence is that it is an intelligence (Mayer et al. 1999). As such, emotional intelligence can be improved through maturity and training. Research completed by Fabio Sala (2002) supports this statement. The results of Sala's studies demonstrate a marked improvement in emotional intelligence of individuals who participated in emotional intelligence training programs (Sala 2002).

The authors build upon the current research domain of emotional intelligence. In particular, they suggest understanding project participants' levels of "soft" skills and abilities (emotional intelligence) will provide a better understanding of the challenges toward developing the next generation of leaders. Additionally, to effectively train an individual, it is important to understand how the training should be focused. To achieve this, it is important to evaluate the level of the individual's emotional intelligence and expose the individual's strengths and weaknesses. The following section provides an assessment of an emotional intelligence inventory directed toward construction industry leaders.

An assessment of construction industry executive behavior

The authors previously investigated the emotional intelligence of construction industry executive leaders. The results of the study provide the motivation for pursuing the question of what type of behavioral traits current industry executives possess in relation to the challenges and required attributes identified (Songer and Chinowsky 2008). Additionally, this comparison provides the basis for determining the gaps between behavioral traits and anticipated challenges. The study results are summarized here.

The summary below identifies emotional strengths and weaknesses of construction industry leaders. Additionally, the relationship between leadership challenges and emotional intelligence (EI) is discussed.

EI strengths

The top three emotional strengths among construction leaders assessed were stress tolerance, independence, and optimism. Stress tolerance falls under the stress management area of EI, independence is categorized under intrapersonal skills and optimism is a component of general mood.

Stress tolerance

A common construction industry understanding is that there is stress associated with the majority of jobs in the industry. There is continued pressure to complete projects within time and budget constraints. Additionally, there is risk involved in undertaking a construction project because of the uniqueness and uncontrolled nature of the sites. Also there is financial pressure because profit margins are historically low, reducing flexibility to make mistakes and still gain financially. The leaders assessed in this study tend to be resourceful and effective as a way to handle stressful situations. The data demonstrates they are well versed in developing suitable methods to deal with adversity and stress. Additionally, the participants demonstrate an overall belief in their own ability to face and handle stressful situations.

Independence

The next strongest EI subscale for the executive group was independence. Scoring high in the area of independence means the group has a high degree of self-confidence as well as the ability to make decisions. As a whole the leaders surveyed have the ability to think independently as well as involve others in their decision-making process. These are undoubtedly characteristics strong construction leaders need to be successful. The higher score in independence illustrates that the leaders surveyed possess these characteristics and are in high positions within their companies.

Optimism

Finally, optimism was found to be a strong EI component of the executive group assessed. This characteristic measures the way the group regards the future. This subscale is subject to market conditions.

EI *weaknesses*

The EI weaknesses of construction industry executives are each components of the Interpersonal skills. The three weakest subscales for the group were: empathy, interpersonal relationship and social responsibility.

Empathy

Empathy is an area where women strongly outscored the men in the study. The ability to be aware of, to understand, and to appreciate the feelings of others is not necessarily something practiced often in the construction industry. Historically, construction leaders did not need to be aware of others' feelings because it was an industry of "low bidder wins". There were no hurt feelings if one subcontractor was chosen over another because of a lower price.

Interpersonal relationship

The second weakest EI area for the group was interpersonal relationship. This subscale aligns with empathy in that it is the ability to form intimate relationships with others. Again, this was an area where the women studied outscored the men by a large margin. Construction is not considered a "touchy-feely" industry by any means. As mentioned above, relationships with subcontractors were historically based on pricing, not how well one "liked" the subcontractor. Construction leaders did not need this aspect when they were expected to act as tyrant-type rulers. Intimate comfortable relationships were not typically formed between construction leader and subordinate.

Leadership challenge–EI relationship

The EI weaknesses have a direct relationship to the required leadership desired traits identified by current industry executives. Specifically, the need for interpersonal skills and communication are of primary importance to the next generation of leaders. Closely aligned with those traits are the needs to be visionary and advocate change. Each of these traits is dominated by the ability to communicate and develop personal relationships with organization members. This emphasis on communication and relationships raises a red flag due to the fact that the primary EI weaknesses found in industry executives are in interpersonal skills. The specific EQ(Emotional Quotient)–leadership gaps include visioning, interpersonal skills, decision making, and independence.

EQ researchers assert that higher EQ scores relate to higher levels of creativity in individuals. This connection is relevant to vision based on the fact that developing a vision is a creativity-based task (Chinowsky and Meredith 2000). Specifically, the ability to establish visionary goals requires an individual to break from the pragmatic bounds of traditional project management. However, the total EQ scores evident in this study illustrate that the scores are only average against the general population. Therefore, visioning ability is an area of leadership development requiring focus if organizations intend to incorporate goals that extend beyond the bounds of traditional objectives.

In the most direct relationship detected in the prior research, the establishment of interpersonal skills as a primary requirement for executives is directly opposed to the finding that construction executives have a primary EQ weakness in interpersonal relationships. The gap in this relationship is arguably the most significant for the construction industry. This area has been downplayed for many years as less important than technical skills, but it is now apparent that this area requires additional attention by the industry as a whole and within each organization specifically.

Construction leaders display a strong independence trait in their personalities. This is expected because running a project at a remote project site requires individuals to take responsibility for actions and be independent in their decision making. However, this strength can become a potential weakness at a leadership position. Specifically, running an organization requires individuals to consider the advice of experts within the organization from areas such as finance, project development, and project management. If an executive relies too heavily on an independent trait, then this advice may be downplayed and critical decisions could be made with incomplete data. Therefore, an additional challenge identified in this study is a focus on how to encourage senior leaders to reduce their independence and take other opinions into consideration.

These gaps or potential weaknesses highlight the potential difficulties that may arise in organizational management if EQ attributes are not considered in the development of future leaders. The challenge then becomes developing appropriate leadership development initiates. It is clear that current executives recognize the need for individuals to possess skills in the area identified. However, it is also clear that a weakness exists in these skills. Thus, there is an identified need to change the path through which the next generation of industry leaders are prepared for their positions. Specifically, a greater emphasis on interpersonal training is required to meet the identified traits. If changed emphasis does not occur, the future generation of leaders may not have the abilities to address issues such as workforce, globalization, and attracting qualified professionals that have been identified as industry challenges. The evolution of collaborative project delivery in the construction industry reinforces this concept of leader as integrator. The following sections expand upon this relationship of emerging integrated delivery with leadership requirements.

Leading collaborative efforts in construction

Project delivery within the construction industry continues to evolve toward increasingly collaborative environments. Current terminology for collaborative project delivery efforts include alliancing, relational contracting, and integrated project delivery. Common among all collaborative efforts is the documented need for improved communication and trust among project participants. Successful transformation to collaborative delivery requires fundamental shifts in organizational operations and attendant culture. Project team development, project processes and the underlying norms, beliefs, and attitudes must not continue to be based in traditional performance-only construction environments. Sustainable collaboration requires additional emphasis on social aspects throughout the project lifecycle. Establishing collaborative cultures among construction project participants requires a heightened sense of communication and trust when participating in integrated project delivery. The next generation construction leader must develop the skills necessary for transforming traditional corporate cultures of isolation into those of true integration.

Globally, there has been a call for radical and dramatic "cultural" change in the construction industry recommending increasing collaboration through various team working approaches such as partnering, project alliancing, relational contracting and integrated project delivery (IPD). Hong Kong-based studies on joint risk management through relational contracting (RC) and project alliancing efforts in Australia are examples that reflect a growing desire for cultural change within the construction industry itself. More recently, IPD efforts in the US continue to suggest the value and necessity of collaboration (Martin and Songer 2004). Yet, the fragmented nature of construction continues to challenge the ability for construction to become a truly collaborative process (Rowings, *et al.* 1996).

Collaboration on construction projects requires participants to more broadly understand each other with regard to specific technologies, finances, and operations of the respective participants (Ibbs *et al.* 2003; Martin and Songer 2004). Sharing this information demands a significant level of trust and rich communication among the project participants (Malik *et al.* 2007). Collaborative environments necessarily require a project participant to look beyond self-interest alone and to consider other stakeholders involved with the project (Shin 2001). Prior research by the authors demonstrated the successful implementation of integrated project delivery efforts in construction. This prior research identified the need for fundamental changes to the organizational structure, administrative procedures, design processes, field operations, and the effective use of technology for implementing sustainable collaboration in project delivery.

While specific organizational structures vary among IPD implementations, there is a common theme among IPD projects. There is organization boundary collapse. The individual yet homogenous organizations unite and assume responsibility for their individual scopes of work and the overall performance of the project. This may manifest as cosignators to the primary contract with the owner, coindemnitors to the project performance bond, or some other expressed commitment to each of the collaborating organizations.

This project-centric commitment results in changes to traditional administrative procedures. In some cases, information sharing and financial responsibility are integrated and indentified through contract. In a more truly integrated effort, each member of the IPD team is proportional owner of the joint company and all costs and profits are shared as well as planning, budgeting, and decision making. Each member is reimbursed for direct job costs at the end of each month and then at the end of the project, the gross profit is distributed amongst the primary team members. Discussions with a variety of IPD participant organizations verifies that some form of direct financial investment in the project creates the necessary incentive for responsive collaboration.

In IPD, planning and coordination of the work becomes the responsibility of each organizational member. To a large extent, this is relegated to the people who perform the work to plan and coordinate amongst themselves. Budgets are established by combining individual estimates from each portion of work. The combining of labor, material, and equipment costs create the unified budget identifying potential duplications and gaps. Because of the mutual ownership of overruns and profits, constructive decisions are generally answered in the field. Problems are identified, solved, and then documented instead of the traditional identified, documented, negotiated, and then solved. There is no need for change orders within the team or back-charges as any additional cost is allocated proportionately amongst the members.

The IPD design process facilitates designer-constructor integration. The contractors that have expertise in highly technical areas tend to be the most involved in the collaborative design process. In concert with the designer, the entire team develops and analyzes possible alternatives. The optimal design is selected and budgeting is then determined. In the fully integrated delivery approach, contractors work with designers in producing the project documents and sign off on every design produced. Therefore, constructability is maximized. In short, the project documents become fabrication drawings rather than engineering drawings. Designers, engineers, and construction personnel are located on-site to provide immediate response to questions and concerns and to develop coordination documents as necessary.

Fully integrated project delivery provides tremendous opportunities in modifying the traditionally decentralized field procedures in construction. Specialty trades in an IPD arrangement are encouraged to take on activities that are outside their original scope of work (provided no safety precautions are violated) in order to streamline construction and make succeeding work easier for other team members. For example, if the plumbing contractor can spend an additional $5,000 to save the electrical contractor $20,000, the overall project (and its IPD participants) benefit. A significant factor contributing to the optimization of field procedures is the opportunity of composite teams. For example, during the renovation of a dormitory, a fire sprinkler system, central hot water, data terminals, cable TV, and phone systems for each room were required to be added to the contract. Instead of having each subcontractor perform all the tasks separately, a composite crew developed to complete each room greatly enhances efficiency.

The IPD project team utilizes technology to reduce schedule and costs. Global website technology is used to provide critical information to all team members, including the owner. Drawings, budget changes, schedule impacts, key coordination issues, and value engineering ideas are all posted to the website. This allows for timely identification and resolution of design and construction issues by the appropriate individuals that are equally informed.

Thus, changes in organizational structure, administrative procedures, design processes, field operations, and the effective use of technology are necessary for implementing sustainable collaboration in project delivery.

As described, the fundamental element for achieving Integrated Project Delivery is rich collaboration. This collaboration requires a free exchange of knowledge based on trust and communication. However, achieving collaboration is the challenge that divides organizations between those that are able to deliver IPD and those that remain a single discipline specialist. The authors developed an Integration Matrix, which provides a visualization of this difference based on professional trust and client-specific communication (Songer *et al.* 2009). These two factors form the foundation for achieving collaboration and thus represent the two axes on which the Integration Matrix measures achievement towards IPD.

With professional trust and specific information as the key measurement factors, the Integration Matrix is divided into four quadrants based on the density measurement within each of these networks for a specific project or organization. The underlying premise of the matrix is that the higher the density in these networks, the greater the potential exists to achieve knowledge exchange and collaboration. As illustrated in Figure 2.2, the Integration Matrix is divided into four quadrants. These include; isolation, fragmentation, connection, and integration.

Isolation

When little trust or communication is evident in a team, the likelihood for collaboration is very low. The combination of low density factors in these networks creates a situation where individuals focus on individual tasks and little interaction is occurring on the team. In this state of isolation, the team is limited in its potential due to an absence of interaction or belief that benefit will occur through interaction.

Fragmentation

When communication is increased within a team, but trust remains low, a state of fragmentation emerges within the organization. In this state, groups of individuals communicate information, but the lack of trust inhibits this information flow from developing into collaboration. This is the most common of the project states as it reflects the fragmentation within the AEC industry where contractual and professional issues contribute to individual specialties and lack of trust between project stakeholders.

Connection

The opposite of the *fragmentation* state is the *connection* state. In this state, trust is evident within the team, but communication density is low. The impact of this combination is to create a team where enthusiasm and interest may exist to collaborate, but the communication links are not established to put in place the interaction. Thus, the intent exists, but organizational barriers inhibit the intent from becoming realized.

Integration

The preferred state within the Integration Matrix is integration, that is to say true collaboration. In this state, both the professional trust network and the specific communication network indicate high-density levels that translate into the requirements to achieve collaboration. In this state, the network individuals indicate that they both trust the other team members and have a high level of communication within the team. This

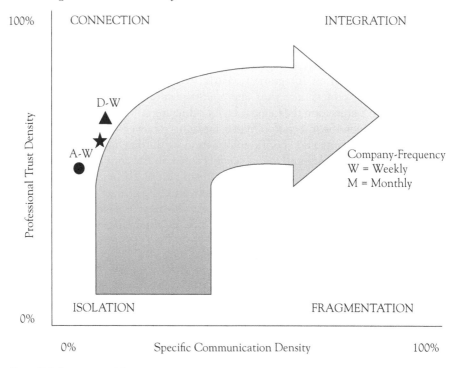

Figure 2.2 Integration Matrix.

integrative state is the fundamental requirement for transforming the team from individual stakeholders to a collaborative group that can effectively deliver an IPD project.

The density of the professional trust network together with the density of the specific information exchange network places an organization into one of the four states within the matrix. The desired state for an organization pursuing IPD is the upper right of the matrix, integration. The relationship between the Integration State and IPD is based on the need to move toward an effective, collaborative environment in which a team operates. If a team has low trust within the network, then two critical barriers emerge within the team environment. First, the lack of trust creates a contentious rather than a collaborative environment. In this environment, legal, functional, and contractual roles supersede integration as the guiding principles in team operation. Second, the lack of trust places an overemphasis on information transfer within the operating environment. In this environment, individuals focus extensively on information transfer guidelines from items such as schedules or work breakdown structures. This emphasis leads to a focus on efficiency of information transfer rather than effectiveness of team interaction.

The Integration Matrix illustrates the impact of the traditional industry fragmentation on the ability to deliver an IPD project. The discipline barriers that are erected within an organization or between organizations inhibit both communication and professional trust from developing at a rate that is conducive to IPD. From a leadership perspective, the challenge is clearly illustrated. The organization leadership needs to refocus on the elements outlined at the beginning of this chapter to encourage participants to build trust and communication to move the organization into the Integration sector.

Transformative strategies for the collaborative leader

If a team is diagnosed as currently exhibiting high trust, but low communication (Quadrant II: Fragmentation) key leadership strategies for moving toward Quadrant IV: Integration would include:

- If team members are "working at a distance," leaders can orchestrate forums for the team to spend more time together face to face. These can include both work sessions and opportunities for social events. A more dramatic strategy used by some IPD teams is to physically collocate team members from each of the core project organizations or groups in the same space.
- If the principal barrier is "communication style" (for example poor communication resulting from introverted personality types) leaders can engage the team in learning processes to understand their individual styles better, how to connect to and leverage each others' capacities.
- If the problem is "conflict aversion"(limited communication due to discomfort in dealing with conflict around differences) leaders can teach the team tools for engaging in constructive conflict and can facilitate crucial conversations among team members. Over time teams build these muscles and learn that conflict, when handled constructively, can be a significant source of creativity and connection. If necessary, outside facilitators can be brought in to assist teams in resolving current conflicts and building the capacity to engage in productive conflict on their own.
- If the problem is "lack of commitment to the team" leaders can coach participants in how to make and follow through on reliable promises involving group participation.

If a team is diagnosed as currently exhibiting low trust, but high communication (Quadrant III: Connected) key leadership strategies for moving toward Quadrant IV: Integration would include:

- Spend time together doing real work that allows for the expertise of individuals to be visible. Leaders can facilitate group planning, design, problem solving, and decision-making processes followed by reflection and learning about the effectiveness in completing these efforts.
- Visibly recognize and celebrate both individuals and groups as they demonstrate their expertise and reliability in successfully following through on commitments.
- Help the team move beyond superficial, procedural communication (exchange of technical information) to a higher degree of intimacy about the matters at hand, surfacing personal thoughts and feelings about the project. Leaders can model this behavior by demonstrating their own willingness to share personal perspectives and be vulnerable. They can also orchestrate team-building exercises that help people learn more about each other and engage them in deeper conversations that disclose personal perspectives or concerns.
- Surface and explore systems archetypes that may be contributing to a lack of trust and below the line behaviors (blaming, playing the victim, etc.). Systems thinking teaches that problems (that are contributing to distrust) most often are not the fault of particular individuals, but rather are present in the structure of the current system people are part of. Leaders can reframe problems that are undermining trust in this way and help the team understand how the system is causing these problems and identify leverage points

that can be used to change adverse situations.
- Act on instances where individuals exhibit a lack of trust and respect. Engage in constructive conversations with the people involved to surface assumptions and mental models driving that behavior and give those individuals the opportunity to change their beliefs. If they are unwilling to trust and live according to the team's creed and values individuals should be asked to leave the team.

Conclusion

The authors put forth their view on leadership by focusing on the building of leaders through incremental steps of increased responsibility. Rather than initially focusing almost exclusively on technical responsibilities, it is important to start preparing future leaders from the early stages of their careers through exposure to organizational concerns and individual responsibilities (integrator model). From this perspective, the concept of leadership focuses on leaders understanding and embracing the need for improved emotional competencies and collaborative ability.

Key items for putting leadership into practice

Changing the long-held beliefs about what is appropriate for industry leadership preparation is something that will be difficult for many organizations to achieve. However, to assist in the transformation process, several key items should be considered as focal points for the organization.

- **Awareness of issue** – The first step in the process is to ensure that the organization is aware of the new leadership demands. The organization should commence a task force to examine the new needs and their relevance to the organization.
- **Broadening of definition** – Once awareness exists, the next step is to broaden the definition of what are key leadership attributes. It is time to reduce the focus on technical ability and broaden the concept of leadership.
- **Leadership responsibilities** – Leaders need to establish vision. Leaders need to be visionary. Leaders are not being productive if they are micro-managing. Redefine the responsibilities of the leader to focus on leadership.
- **Focus on education** – Developing leadership skills requires many years of development. An organization must realize that this education is an investment in the future and place it as a core requirement for long-term success.
- **Opportunity** – Leaders need to be gradually introduced to the requirements of organization leadership. The successful introduction of these concerns requires a gradual introduction to the needs of the organization. Allow future leaders to gradually experience the demands of leadership.
- **Breadth not depth** – The final item for all organizations to understand is that the future leader will have a breadth of knowledge, but not necessarily be the most in-depth expert on a topic. Understanding, acknowledging, and adopting this philosophy is the ultimate key to moving the organization in this direction.

References

Ashforth, B.E. and Humphrey R.H. (1995), "Emotion in the workplace: A reappraisal," *Human Relations*, 48(2): 97–125.

Black, C., Akintoye, A., and Fitzgerald, E. (2000), "An analysis of success factors and benefits of partnering in construction," *International Journal of Project Management*, 18(6): 423–434.

Bogus, S.M. and Rounds, S.L. (2006), "Incorporating leadership skill development in construction training programs," *2nd Specialty Conference on Leadership and Management in Construction*, Construction Research Council, ASCE-CIB, Grand Bahama Island, Bahamas, May 4–6.

Bresnen, M. and Marshall, N. (2000a), "Partnering in construction: a critical review of issues, problems and dilemmas," *Construction Management and Economics*, 18(2): 229–237.

—— and —— (2000b), "Motivation, commitment and the use of incentives in partnerships and alliances," *Construction Management and Economics*, 18(5): 587–598.

Campion, M.A., Papper, E.M., and Medsker, G.J. (1996), "Relations between work team characteristics and effectiveness: a replication and extension," *Personnel Psychology*, 49(2): 429–453.

Carr, P.G., de la Garza, J.M., and Vorster, M.C. (2002), "Relationship between personality traits and performance for engineering and architectural professional providing design services," *Journal of Management in Engineering*, 18(4): 158–166.

Cheng, E.W.L. and Li, H. (2001), "Development of a conceptual model of construction partnering," *Engineering Construction and Architectural Management*, 8(4): 292–303.

Chinowsky, P.S. and Meredith, J.E. (2000), *Strategic Corporate Management for Engineering*, New York: Oxford University Press.

Gibson, J.L., Ivancevich, J.M. and Donnelly, J. (2003), *Organizations: Behavior, Structure, Processes*, New York: McGraw-Hill/Irwin.

Goleman, D. (1998), *Working with Emotional Intelligence*, New York: Bantam Dell.

——, Boyatzis, R., and McKee, A.(2001), "Primal leadership: The hidden driver of great performance," *Harvard Business Review*, 79(11): 42–51.

Ibbs, W.C., Kwak, Y. H., Ng, T., and Odabasi, A.M. (2003), "Project delivery systems and project change: Quantitative analysis," *Journal of Construction Engineering and Management*, 129(4): 382–387.

Johnson, H.M. and Singh, A. (1998), "The personality of civil engineers," *Journal of Management in Engineering*, 14(4): 45–56.

Kemper, C.L. (1999), "EQ vs. IQ – emotional intelligence, intelligence quotient," *Communication World* 16(October): 15–22.

Kouzes, J.M. and Posner, B. (2002), *The Leadership Challenge*, San Francisco, CA: Jossey-Bass.

Ling, F.Y.-Y. (2002), "Model for predicting performance of architects and engineers," *Journal of Construction Engineering and Management*, 128(5): 446–455.

Loosemore, M. (1998), "The methodological challenges posed by the confrontational nature of the construction industry," *Engineering Construction and Architectural Management*, 5(3): 285–293.

Malik, M.A., Khalfan, P., and McDermott, W.S. (2007) "Building trust in construction projects," *Supply Chain Management: An International Journal*, 12(6): 385–391

Martin, D.W. and Songer, A.D. (2004), "Contracts vs. covenants in integrated project delivery systems," *Construction Information Quarterly*, CIOB, 6(2): 51–55.

Mayer, J. D., Caruso, D.R., and Salovey, P. (1999), "Emotional intelligence meets traditional standards for intelligence," *Intelligence*, 27(4): 267–298.

Mayer, J.D., Salovey, P., and D.R. Caruso (2000), "Models of emotional intelligence," in R.J Sternberg (ed.), *Handbook of Intelligence*, New York: Cambridge University Press, pp. 396–420.

Pocock, J. B., Liu, L.Y., and Kim, M.K. (1997), "Impact of management approach on project interaction and performance," *Journal of Construction Engineering and Management*, 123(4): 411–418.

Rowings, J.E., Federle, M.O., and Birkland, S.A. (1996), "Characteristics of the craft workforce," *Journal of Construction Engineering and Management*, ASCE, 122(1): 83–90.

Sala, F. (2002), *Do Programs Designed to Increase Emotional Intelligence at Work - Work?* Boston, MA: Hay/McBer, online, available at: www.eiconsortium.org/pdf/mastering_emotional_intelligence_program_eval.pdf.

Scheusner, H.K. (2002), *Emotional Intelligence Among Leaders and Non-Leaders in Campus Organizations*, master's thesis, Blacksburg, Virginia Polytechnic Institute and State University, online, available at: http://scholar.lib.vt.edu/theses/available/etd-05032002-095612/unrestricted/beginning1.pdf.

Shin, K. (2001), *The Covenantal Interpretation of the Business Corporation*, Lanham, MD: University Press of America.

Singh, A. (2002), "Behavioural perceptions of design and construction engineers," *Engineering Construction and Architectural Management*, 9(2): 66–80.

Skipper, C.O. and Bell, L.C. (2006), "Influences impacting leadership development," *Journal of Management in Engineering*, ASCE, (22)2: 68–74.

Songer, A.D. and Chinowsky, P. (2008), "Leadership behavior in construction executives: Challenges for the next generation," *Construction Information Quarterly*, CIOB, 10(2): 59–66.

——, ——, and Butler, C. (2002), "Construction engineering professional 2020," a report to the National Science Foundation, prepared under contract no. CMS-0240156.

——, ——, and Davy, K. (2009), "Social network analysis for determining integrated project delivery readiness," LEAD Conference, Lake Tahoe, CA, online, available at: http://epossociety.org/LEAD2009/papers.htm.

Thamhain, H.J. (1992), *Engineering Management, Managing Effectively in Technology-Based Organizations*, New York: John Wiley & Sons, Inc.

Thomas, S.R., Tucker, R.T., and Kelly, W.R. (1998), "Critical communications variables," *Journal of Construction Engineering and Management*, 124(1): 58–66.

Williams, W.M. and Sternberg, R.J. (1988), "Group intelligence: Why some groups are better than others," *Intelligence*, 12(4): 351–377.

3 Knowledge management in global environments

Amy Javernick-Will and Timo Hartmann

To remain viable, organizations must manage their resources strategically. This not only includes managing tangible assets, such as money and property, but also intangible assets, such as knowledge. Moreover, management scholars have identified knowledge as one of the key ingredients to sustain competitive advantage, primarily because companies generate increasingly less return on traditional resources, such as labor, land, and capital (Spender 1996). Despite the pressing need to manage knowledge, many organizations "don't know what they know" and all organizations "know more than they can tell" (Spender 1996). If organizations do not have processes for learning new knowledge and sharing existing knowledge, they face the possibility of repeating mistakes, "reinventing the wheel," and wasting valuable time and resources. Organizations must make knowledge available for their employees and support them in developing new knowledge. In this chapter, we focus on outlining a number of strategies that companies can use to do so.

This chapter begins by discussing how knowledge is different from information and identifying characteristics of knowledge that companies must understand to manage knowledge meaningfully in their organizations. Based on these knowledge characteristics, the chapter outlines strategies that companies can implement to enable their employees to combine and share their knowledge strategically. The chapter continues with a discussion on how to implement a knowledge management solution. We then introduce the STEPS method that companies can use to implement a knowledge management strategy progressively in a top-down manner. In addition, the chapter provides a number of recommendations for how to support the knowledge management implementation from the "bottom up"—i.e. from the operational level of the company. The chapter closes with common barriers that companies face during the implementation and how to overcome these barriers.

Knowledge

Before we discuss specific implementation strategies, it is important to understand the definition and characteristics of knowledge. To gain this understanding, it is helpful to distinguish knowledge from information (Brown and Duguid 2001). The main difference between knowledge and information is that knowledge entails somebody who knows. With respect to information we may ask: "Where is this information?" while with respect to knowledge we need to ask "Who knows what?" Because knowledge is a personal characteristic, it is hard to detach from individual employees. For instance, people can collect, transfer, and store information easily. Doing the same with knowledge is difficult. In addition to the personal nature of knowledge, knowledge is also different from information as it presumes a level of understanding. We can say "I got the information, but I do not understand it," while we cannot say "I got the knowledge, but I do not understand it." With this initial understanding of what knowledge is, we introduce a number of knowledge characteristics in the next subsections that can help companies to categorize different types of knowledge that are important for their operations. This categorization is a first step in implementing a knowledge management strategy that enables employees to share knowledge that is important for their organization.

Tacit vs. explicit

On a general level, there are two primary types of knowledge that people can possess. The first kind is knowledge people can talk about freely—they "know that." The second kind of knowledge people possess, but they have a hard time explaining and sharing this knowledge with others—they "know how." Polanyi (1967) labels these two kinds of knowledge explicit knowledge ("know that") and tacit knowledge ("know how").

A classic example of tacit knowledge is riding a bike (Cook and Brown 1999). Even if a person (knowledge transmitter) can ride a bike, she may not be able to communicate this knowledge to another person (knowledge receiver) in a way that this person will be able to ride a bike. The knowledge transmitter can explain how she rides a bike in as much detail as possible and still the knowledge receiver would likely fall during her first few bike-riding trials. The knowledge transmitter's knowledge about *how to* ride a bike is tacit and the knowledge receiver can often only internalize this knowledge by going through the tedious procedure of practicing bike riding.

There are many examples of tacit knowledge in construction engineering and management practice. For example, experienced cost engineers in many companies hold tacit knowledge regarding estimating contingencies. During the cost estimating process, cost engineers use a great deal of explicit knowledge to predict the costs of a project: They take off quantities of all building elements from construction drawings and specifications. Then they use historical data from their company's database to determine the construction costs for each of these elements. Such construction drawings and historical company databases allow estimators to transfer knowledge about costs or scope of a project easily between several estimators and even from project to project—the knowledge is explicit. Estimators *know that* on average, a particular crew will require a specific duration to build one unit of a specific building component. After determining the price for each of the components of the building, cost engineers need to determine contingencies for each of the previously defined component costs to adjust the costs to the local project context. In many cases, explicit knowledge about how to define these contingencies does not exist. Knowledge about contingencies is complex and depends on many closely interrelated factors. Prices for material vary from season to season, region-to-region, or even according to certain unpredictable market conditions. Similarly, the productivity of workers will vary from project to project. Many estimators believe that the "real art of their profession" is accounting for these project specific differences. To determine realistic contingencies, estimators need a great deal of practical estimating experience. If asked, many experienced estimators will not be able to communicate explicitly *how* they chose a certain contingency value for a specific project. Therefore, the knowledge regarding estimating contingencies is tacit. It is deeply integrated in estimating practice, complex, and, largely, context specific.

Complexity

The more complex the knowledge, the more difficult it is to transfer. It can take time to convey complicated concepts and thus requires more attention and thought regarding processes to transfer the knowledge. In some cases, complex knowledge can take years to acquire, understand and put into practice. Because complex knowledge is so hard to accrue people often relate it to tacit knowledge. Examples of complex tacit knowledge include the skill of the experienced machinist, who knows that a particular noise from a piece of equipment means that they need to fix a specific component; or the ability of an experienced real estate developer to "read" the local real estate market to determine the appropriate building type to build in a certain location. But complex knowledge can also be explicit knowledge that is externalized in formulas, and presented in textbooks. It is, for example, not easy for non-structural engineers to acquire the vast amount of structural engineering knowledge that is available explicitly in the form of building codes to calculate whether a specific building fulfills local structural regulations. Tacit or explicit, complex knowledge requires years of training and background before employees can understand it.

Observability

If knowledge is hard to observe, it will be more difficult to transfer. Observability describes the ease by which other employees in the company can gain an understanding of personal or organizational knowledge by watching how other employees or organizations conduct a task by applying their knowledge. Observability of knowledge is, therefore, simply the ease by which employees in the company can observe and detect knowledge that others possess. Again observability of knowledge is independent from its tacit or explicit character. On the one hand, it is easy to observe somebody riding a bike; however, the bike rider will have a hard time explaining how to ride a bike. On the other hand, it is difficult to observe the process of how estimators put together cost proposals without asking them in detail what they do. Once asked, however, the estimator can easily explain the steps he followed during the quantity take off task. In a business context, it is important for firms to make their work processes easily observable by members of the company to foster company internal learning. At the same time, companies want to make the knowledge that provides them with a strategic advantage, such as processes, hard to observe by their competitors. This makes strategy regarding observability particularly challenging for organizations in highly competitive fields.

Context-specific/institutional

Some knowledge is very specific to and embedded within a particular context. For instance, certain engineering principles are more applicable to different geographical regions, such as geotechnical or seismic design. Construction techniques can also vary based on location—for instance, building in Russia will require different techniques and scheduling than building in West Africa. These are types of institutional knowledge—regulations, norms, and culture—which will vary depending on the location of the project. The types of institutional knowledge type that are important to an organization will vary based upon firm type (Javernick-Will and Scott 2010) and will require different processes to transfer the knowledge depending on institutional knowledge type (regulative, normative, or cultural-cognitive (Javernick-Will and Levitt 2010). Context-specific, or institutional knowledge, is explicit or tacit. For example, in many regions, written regulations exist that allow employees to search for applicable laws for their projects. Such knowledge is available on external websites or in books. After collecting this knowledge for projects, employees can document the knowledge on their organization's intranet and indicate the region(s) that the laws apply to. Context-specific knowledge can also be tacit. For instance, social norms and gestures or cultural beliefs are engrained into people's everyday lives and are often taken for granted. In addition, in many countries, processes and procedures, such as obtaining a permit or buying land, are tacit and not available in written form. In these instances, it is often easiest to identify a person within or outside the company that has experience or a background in the area and holds the knowledge that can help to train others and provide the contextualized knowledge.

Half-life

The half-life of knowledge—the length of time before the value of knowledge to the organization declines to 50 percent of its original value—can vary enormously from industry to industry (Haindl 2002). For instance, people widely agree on certain scientific principles that were conceived centuries ago. However, other types of knowledge become obsolete quickly. In the IT sector, for instance, there is general acceptance of a half-life of approximately one year (Haindl 2002). In comparison, many structural engineering principles have been in place for decades. The length of time knowledge is valid should influence an organization's knowledge management strategy as the organization will need to consider the resources required to transfer the knowledge as well as the processes required to keep the knowledge up to date (Javernick-Will and Levitt 2009).

Strategies for knowledge management

Once companies have identified the types of knowledge that are important for their operations, they can start to implement a knowledge management strategy that enables the management of the identified knowledge categories. Companies that wish to implement a knowledge management strategy need to answer two major questions:

1   How can we support our employees to create new practice relevant knowledge and, in this way, continuously improve best practice?
2   How can we diffuse existing knowledge through the company to transfer and align best practices?

Methods and strategies exist that companies can implement to support their employees with knowledge exploration, i.e. how to create new knowledge, and exploitation, i.e. how to diffuse existing knowledge through the company.

Knowledge exploration

In today's ever-changing marketplace, organizations must be able to learn from internal and external sources to adapt to new environments and sustain continuous improvement and competitiveness. A learning organization (Garvin 1993) is an organization whose employees, individually and collectively, continuously develop their capability to create attractive outcomes for the organization.

Kululanga *et al.* (1999) identified learning mechanisms employed to acquire knowledge for continuous improvement. These included:

- **Collaborative arrangements** where firms learn from others in partnerships, including partnering, alliancing, joint-venturing, subcontracting, and other agreements;
- **Noncollaborative arrangements** where firms learn via acquisitions and mergers;
- **Networks**, where firms learn from other institutions or professionally based social networks;
- **In-house learning** where firms learn through research projects, reviews, benchmarking, and trial projects;
- **Individual learning efforts** during which firms try to increase the knowledge of individual employees through seminars, or training; and
- **Hiring strategies** by attracting new employees who already possess specific knowledge.

Organizations must proactively plan and engage their employees in learning processes to gain new knowledge and to keep the company's knowledgebase current to maintain a competitive advantage. Once new knowledge is gained and assimilated, companies can then leverage this knowledge across the organization by focusing on knowledge transfer.

Knowledge exploitation

For knowledge to flow through an organization, employees in different project teams and geographical locations must share it. Nonaka's theory of knowledge conversion (Nonaka 1994; Nonaka and Takeuchi 1995) describes how employees can convert and share their tacit and explicit knowledge with others. According to Nonaka, employees can transfer explicit knowledge directly through the process of "combination". Further, it is also possible to transfer tacit knowledge directly through "Socialization," for instance, through mentoring programs, or on-the-job training. However, often the direct transfer of tacit knowledge is not an easy task. Therefore, employees often transfer tacit knowledge indirectly by converting their tacit knowledge into explicit knowledge first ("externalization"). They can achieve this, by reconstructing how they routinely do something. Once they have transferred their tacit knowledge into explicit knowledge, they can then transfer this explicit knowledge to another person with "combination". This person then transfers the explicit knowledge back into tacit knowledge by applying the explicit knowledge in practice. Nonaka calls this final transformation process internalizing the knowledge ("internalization"). Table 3.1 summarizes Nonaka's four different knowledge transformation and transfer processes. In the next subsection, we explain each of the above knowledge transfer mechanisms in more detail and discuss methods of how companies can support their employees with each of these ways to transfer knowledge.

Transfer of explicit knowledge—combination

The transfer of knowledge through *combination* relies heavily on transferring knowledge in written form. Companies can employ procedures, manuals, standards, schedules, check-lists, or templates to integrate and share explicit knowledge among their employees. This practice of combination is closely linked to information and communication technology (ICT). ICT helps to aid the formal integration of written, explicit knowledge by simplifying the process of assimilating, storing and retrieving {Rockart and Short 1989}. In fact, many people

Table 3.1 Knowledge transfer mechanisms

	EXPLICIT (Know That)	TACIT (Know How)
EXPLICIT (Know That) ->	Combination	Externalization
TACIT (Know How) ->	Internalization	Socialization

Source: Nonaka and Takeuchi (1995).

automatically associate knowledge management with building information databases that enable the transfer of explicit knowledge systematically and routinely. (Ball 2006)

Javernick-Will and Levitt (2010) identified four primary combination methods that global organizations within the architecture-engineering-construction (AEC) community use to transfer knowledge: project databases, reports, procedures and processes, and a general-purpose online system such as a corporate Intranet.

- **Project databases** contain codified knowledge that typically provide comparable details or statistics of past projects. These databases can contain project information such as location, client, sector, project value, requests for information, financial transactions, or subcontractors. Employees can access these databases to help prepare proposals and estimates for similar projects and retrieve information regarding contracts, subcontractors, or employees who worked on similar projects in the past.
- **Reports** also help to disseminate written information from projects. This can include executive briefing reports, internal publications that preserve knowledge from the company's experience, and one of the most familiar processes of transferring knowledge within the construction industry, post-project "lessons learned" or "close out" reports. These reports are typically the result of meetings where project staff record their experiences from the project and provide written advice for future project teams. These can include a written account of what occurred at various project stages to maintain a historical record of why project managers made certain decisions and what processes they followed. Unfortunately, however, many "lessons learned" reports have a negative connotation and employees are hesitant to publish their "mistakes" across the organization.
- **Procedures and processes** are formalized procedures that companies request or require their employees to use. In many cases, these procedures and processes are the outcome of reviewing lessons learned across multiple projects. These can include standardized project checklists, such as risk checklists for projects, or processes that companies require their employees to complete during each project stage.
- **General intranet systems** can contain company information and provide links to external Internet sites for things such as building codes, sustainability information, and cost indexes. In addition, these systems can contain plans and specifications of past and current projects.

To add value by combination, companies need to keep the stored explicit knowledge accurate and up-to-date. If companies do not update knowledge regularly, the knowledge may become irrelevant. Such irrelevant knowledge will, in turn, cause frustration and a lack of trust in formal combination processes among employees. This holds especially true for knowledge with a short half-life. In addition, employees must be aware of the knowledge and have access to the stored and codified knowledge. Unfortunately, many organizations store knowledge through combination strategies in a manner where employees do not know that the knowledge exists or are not granted access to it. In this case, the knowledge becomes lost. Employees must be educated with how to find specific knowledge, or, if the company uses an IT solution to combine knowledge, they must have a robust search feature for employees to locate relevant knowledge.

In summary, companies can use combination strategies to transfer explicit knowledge directly among their employees. The overall process required by a combination strategy can also be a valuable tool to transfer tacit knowledge into explicit knowledge. The next subsection describes externalization and internalization strategies to enable such an indirect transfer of tacit knowledge employing combination.

Indirect transfer of tacit knowledge—externalization and internalization

One of the added benefits of the formal combination strategies, described above, is that it requires employees to *externalize* their tacit knowledge into an explicit form. It forces them to externalize relevant tacit knowledge, previously stored in their heads, into a report, database, procedure, or an online system. Through this process, the employees must realize and formalize their knowledge to make it available across the organization. For example, many companies require their project teams to provide a written account of projects and experiences—a "project close out report"— in order to avoid repeated mistakes and share best practices. To generate such reports, project teams have to externalize project facts, decisions, processes, or lessons learned that they had internalized.

After externalization, it is equally important for organizations to emphasize the steps of combination and internalization. Firms often help the project teams to write the project close out reports, but do not require that project teams share the lessons from these reports across the organization. Our research (Javernick-Will and Levitt 2010) at one global contractor showed that most of the employees did not know where to locate the reports or how to determine if "lessons learned" from other projects were applicable to their current assignment. It is important to realize that the indirect transfer of tacit knowledge only works if sound knowledge combination strategies, such as the ones we described in the previous section, are in place.

Additionally, it is also important that companies support their employees with the *internalization* of knowledge. One global contractor supports its employees with the internalization of the knowledge that others had externalized in project close out reports (Javernick-Will 2010). The company required each functional leader to go through all project close out reports to understand and document common findings across projects. The functional leaders then met as a group to compare the findings and discovered similarities across functions. After realizing that many functions had similar issues to correct, they prioritized the list of findings and created training modules and updated procedures and processes to transfer the knowledge throughout the organization. This helped to make employees aware of lessons others learned on previous projects and helped the contractor to avoid common mistakes in the future. Because they focused on specific functions in the organization, this strategy also enabled employees to better internalize the knowledge by making the employees aware of the knowledge, enabling them to actively use the knowledge gained by others on their projects and therefore transferring it into tacit knowledge.

Direct transfer of tacit knowledge—socialization

Many individuals prefer to seek knowledge from individual experts on a personal basis to exchange tacit knowledge directly through *socialization* (Nonaka 1994). Perhaps this is because people are uniquely able to contextualize the knowledge to specific situations and the requestor's experience. A study of six firms within the US AEC sector found that only one company chose an IT-centric strategy (codification); whereas the other five companies relied on a socialization strategy for managing knowledge flow (Carrillo and Chinowsky 2006). Likewise, a study of knowledge-sharing practices in tendering departments in Hong Kong and the UK found that most of the knowledge sharing processes were based on socialization strategies (Fong and Chu 2006). To account for these needs, a knowledge management strategy that uses combination alone is often insufficient and requires the transfer of tacit knowledge directly through socialization strategies, such as mentoring, training, or mimicry (Nissen 2006; Polanyi 1967).

Javernick-Will and Levitt (2010) identified four formal socialization strategies that companies use for transferring tacit knowledge: meetings or teleconferences, reviews, transfer of people between offices, and personal discussions.

- **Meetings/teleconferences** include various forms of meetings between people within the organization. This can include board meetings, asset review meetings, project team meetings, functional meetings, annual conferences, and training seminars.
- **Reviews** consist of a group of employees reviewing investments and assets, operations, audit information, or strategy. Some companies conduct project peer-reviews at various phases, reviewing current plans for design, schedule, budgets, and processes to identify key risks that project teams may have neglected. One of the touted benefits of project peer reviews is the ability of the project team to apply their peer's knowledge immediately.
- Companies can **transfer people** to different divisions and locations with the strategic aim of sharing knowledge. For instance, an organization may transfer a senior employee with knowledge of a particular specialty, such as tunneling, to an office that is expanding its tunneling operations. Likewise, companies may transfer employees that have experience with a particular project type or client to share past knowledge with the new project team. Organizations use this method to transfer company processes and procedures when they are establishing a new office or acquiring another firm. Studies show that deliberately moving members around the organization can be a powerful mechanism to transfer knowledge (Borgatti 2003).
- **Personal discussions** through phone, email, or direct in-person interaction are maybe the most common process to transfer knowledge. These discussions help provide context-rich information tailored to the recipient's current situation and prior experience.
- **Other** socialization strategies include apprenticeship and mentoring programs, training programs, and other groups such as quality circles.

In contrast to combination, socialization allows employees who wish to transfer knowledge to tailor the knowledge to the specific situation and the personal experience of the recipient. In many cases, this allows employees to transfer knowledge that is applicable to a particular project and thus, is easier for the recipient to apply and internalize the knowledge. In addition, many employees feel more comfortable exchanging knowledge, particularly knowledge that was gained from project failures, on a personal level instead of broadcasting it across the organization. Socialization strategies also enable employees to gain intimate knowledge of "who knows what" called referential knowledge. Meeting and interacting with peers helps them to understand each other's past experiences and areas of expertise. This helps them to recall who they should contact when they require particular knowledge.

However, if companies only use a socialization strategy, the ability for the knowledge to reach across the organization is limited. The transfer of knowledge can be restricted to an employee's existing organizational ties with other peers who they have encountered through projects or other interactions. This can limit socialization strategies based upon geographical or functional lines. In addition, employees with considerable experience and knowledge are often highly sought after. The demand for their time can quickly exceed supply (Monteverde and Teece 1982). This can cause frustration for employees who do not have the answer and can cause organizational knowledge to become siloed. Nevertheless, companies can strategically improve existing peer networks to focus on knowledge transfer through socialization as the following example illustrates.

One large design and engineering company organizes a program that enables knowledge sharing across its more than 100 worldwide operating offices by organizing one-week workshops for its recognized experts. The main goal of these workshops is to increase the company's professional network of experts across the organization. To do so, the company uses a strategy to select the participants of the workshops by specifically analyzing the company's social networks. Six months before each workshop, the company sends a survey to each of its recognized technical experts. The survey asks each of the experts to identify the other experts they know on the list. Using the answers from this survey, the company specifically looks for connection gaps in its professional network and invites people to help fill these gaps. During the workshop, the company then organizes breakout work groups with people who do not know each other to specifically establish personal contacts.

Combining combination and socialization strategies through interactive online platforms

Recent knowledge management strategies try to combine socialization with combination processes through interactive online platforms (IOP). A recent study found that three out of five engineering companies used an IOP and one out of four contractors was starting to implement an IOP (Javernick-Will and Levitt 2010). These systems contain static information and explicit knowledge that employees can access, such as processes, procedures and checklists; but they also encourage tacit knowledge exchange through people-to-people connections using forums, searches, and promoting global peer interaction through the organization of communities.

Such communities are groups of employees that share a common function, business or other interest. Some organizations prescribe the communities that employees can join, while others allow employees to form communities based upon interest. Within the communities, interactive online platforms allow employees to post questions via forums. The community leaders and members receive the questions (often through email) and are able to post responses. These can result in a string of responses and questions, resulting in a forum thread. Typically, the questions and responses contain the employee's name and contact information for further follow-up. This helps promote referential knowledge of who knows what within the organization. In addition, most of the interactive online platforms contain robust search features. The search will typically allow employees to search through the explicit knowledge, forums and people in order to find the knowledge they need.

The greatest benefit of IOP is the ability to connect employees and share various types of knowledge across the organization. IOP make it possible for employees to receive advice and experience from peers around the world, extending their limited local social network that exchanges knowledge "around the water cooler" or "the coffee machine." These platforms allow employees to search the entire organization's knowledgebase, not just their office location. In addition, IOP allow new employees that do not have a wealth of prior experience or social networks within an organization to access the organization's knowledge easily. The ability to search through the organization's knowledgebase—whether for explicit knowledge or a person with a particular expertise—decreases queries to their superior, which eases access to the knowledge and decreases the time required to receive a response. If companies keep IOP open for discussion and to members, they can be more democratic, allowing everyone within the organization to provide input by responding to queries on forums. Similar to other knowledge processes, IOP also have limitations. The value of this process will depend on the extent of use across the organization. For instance, if a global company has an IOP and it is only used in North America, the access to and reach of the organization's knowledge will be limited. IOP are based primarily for communication and networking, which helps to keep the knowledge up-to-date. However, if the organization keeps the knowledge and forums online in a static form, the knowledge can become outdated. Companies must therefore ensure that the accessible knowledge is kept up-to-date or has stated limitations. Finally, an organization must decide if it is willing to commit the time and resources to install IOP, train employees to use it, and maintain the platform. The investment required will depend on the size and geographical spread of the organization as well as the types of knowledge they wish to exchange.

One global EPC (engineering, procurement and construction) company has a well-known, award-winning knowledge management program that is based on IOP. The knowledge management team needed a strategy that would enable their offices around the world to access organizational knowledge and communicate with their peers. To accomplish this, they knew that they needed a robust technology that focuses on connecting people. They internally developed the IOP knowledge management system. Many employees tout the benefits of the system, particularly the ability to search the IOP for people, knowledge, and forums. The employees will typically first search for a topic to find related questions and answers. In addition to looking at the responses, they can identify knowledgeable peers by discovering who replies to the questions. If they have a new question, they can post it to the community. Once posters receive a reply, they can obtain additional details by contacting the posting replier directly. In addition, a tremendous benefit of the IOP is the ability to post a question that various time zones can respond to over a 24-hour period. In one example, an employee posted a question in a community forum regarding an economical design solution for a particular operation. Within three days, he received three responses from around the world that provided project references and contacts for different design options. By referencing projects and talking to his global peers, the employee was able to recommend a new concept in the facility design that saved the client approximately one million euros on equipment and operational costs.

Choosing a knowledge management strategy

An organization must decide what it aims to achieve from its knowledge management strategy. A major consideration should include the types of knowledge that are important to the organization. If the knowledge is tacit, it will require an emphasis on *socialization* strategies, or connecting employees. If the knowledge is primarily explicit, the organization can have a greater focus on *combination* strategies that will broadcast the knowledge globally. However, employees must be aware of the knowledge and have access to it for this knowledge management strategy to add value. One of the greatest benefits of IOP is the ability to connect people through a common platform that shares knowledge globally.

If the knowledge is complex, the organization must devote more time and attention to ensuring that the knowledge is transferred to and internalized by the appropriate employees. This may require additional training for specific employees through case studies, simulations, or mentoring to realize the transfer. Knowledge that is hard to observe will require yet another focus of attention. An organization that wishes to manage hard to observe knowledge needs to focus on processes to externalize this knowledge first. However, the organization may wish to protect this knowledge (keeping it hard to observe) so that competitors cannot duplicate their efforts. In this case, they would prefer to keep the knowledge tacit.

Organizations who wish to share context-specific knowledge will need to label the knowledge appropriately to enable meaningful knowledge transfer. IOP, for example, can ensure that transmitters codify their knowledge appropriately according to type of project, project client, or project location so that the recipients are aware of the limitations of the knowledge—for instance, if a particular building technique is only valid in certain climates or conditions, the knowledge would need to be labeled as such to avoid use in a different location. Similarly, organizations may label knowledge for particular project types—i.e. a retail building to provide appropriate context, and some label them based on client type. If the organization relies on socialization strategies, the organization must also train people to state the limitations to the recipient. In many cases, context-specific knowledge will require direct personal interaction to transfer the knowledge (Javernick-Will and Levitt 2010). For instance, cultural knowledge, such as personal gestures, is difficult to transfer through reports. It often requires social interaction to understand the gesture and associated meaning and respond properly. Although employees can use IOP to locate the appropriate person to ask questions of, this type of knowledge still requires personal interaction other than directly through the IOP for transfer. Context-specific institutional knowledge is particularly important for firms operating in multiple markets across the world. Table 3.2 shows the relative frequency of methods global real estate development, contracting, and engineering firms use to share different types of institutional knowledge (all firms) (Javernick-Will and Levitt 2010). As the knowledge becomes more tacit in the cultural-cognitive category, firms tend to rely more frequently on socialization methods to transfer the knowledge.

If the knowledge that is important to an organization has a long half-life, the organization should invest time and resources to make the knowledge explicit and available throughout the organization (Javernick-Will and Levitt 2009). It will thus be more prone to invest in codification strategies or IOP. For instance, engineering firms with a longer half-life of knowledge, such as structural engineers, may be more inclined to invest the resources necessary to make organizational knowledge explicit and available globally than contractors or developers, whose knowledge tends to have a shorter half-life (i.e. the financial market) and be more context-specific depending on location and project.

Table 3.2 Relative frequency of transfer methods for different types of contextual Knowledge across all firms

	TRANSFER METHOD	Contextual knowledge type			
		Regulative	*Normative*	*Cultural cognitive*	*Technical*
COMBINATION	Project database	4.1%	3.8%	3.8%	3.9%
	Written reports	21.6%	18.0%	7.7%	14.5%
	Procedures and processes	10.8%	1.5%	7.7%	7.2%
	General online system	2.7%	3.0%	3.8%	4.6%
	Formal (other)	4.1%	3.8%	3.8%	2.6%
SOCIALIZATION	Interactive online	9.5%	12.8%	0%	25.0%
	Social (other)	5.4%	7.5%	0%	5.3%
	Meetings/teleconferences	5.4%	10.5%	23.1%	6.6%
	Reviews	13.5%	13.5%	7.7%	14.5%
	People transfer	5.4%	6.8%	15.4%	3.3%
	Personal discussions	17.5%	18.8%	27.0%	12.5%
	Column totals	*100%*	*100%*	*100%*	*100%*
	Total references per knowledge type	*n=74*	*n=133*	*n=26*	*n=152*

Implementing a knowledge management strategy

Implementing a viable knowledge management strategy is a lengthy process. The executive management of an organization needs to set a long-term vision about introducing a knowledge sharing culture that increases business performance, and strategically steer the implementation of a knowledge management system to reach that vision. At the same time, the implementation and maintenance of a viable knowledge management strategy depends largely on the buy-in of, and thus, the value the system provides for, employees at the operational level of the organization. Therefore, organizations need to complement their top-down implementation strategy with a bottom-up implementation strategy by working closely together with employees at the operational level during the implementation.

Top-down implementation of knowledge management systems

We suggest that organizations use the STEPS model (see Figure 3.1 opposite) to investigate the relationship between knowledge management and business performance (Robinson et al. 2005) and to support the top down implementation of a knowledge management strategy. The model categorizes different stages of maturity for a knowledge management program. Although originally developed to indicate the maturation stage for an organization's knowledge management program, we believe that reviewing the progressive stages can help

Stage	Primary Activities
Start-Up	Understand the benefits of a KM Solution
Take-off	Strategize a KM solution, including structure, resources, barriers and risks, for your company
Expansion	Expand the awareness of KM throughout the company and address barriers and risks
Progressive	Improve performance of KM through monitoring and measurement
Sustainability	Institutionalize KM with sustained performance of KM activities

Figure 3.1 Activities in the STEPS model (Refer to Robinson et al. 2005 for additional details)

to prepare organizations for the long-term vision and planning that a knowledge management strategy requires. As the program progresses through the stages, the organization will need to focus on new aspects while maintaining their original goals.

The *start-up stage* involves understanding knowledge management concepts, strategies and benefits. In this stage, companies need to determine the need for a knowledge-sharing strategy and the potential that a knowledge management program can have for their organization. The *take-off stage* establishes the goals of the initiative, explores knowledge-sharing process options, commits resources for continued support, and identifies barriers and risks to the strategy. During this stage, organizations should focus on the types of knowledge that are important to the organization to determine their knowledge management strategy. To do this, they should first identify knowledge important to the company and determine the specific characteristics of this knowledge. Based on the identified knowledge characteristics, the company then needs to explore and choose knowledge-sharing processes that are suited to the specific knowledge characteristics identified. Organizations should experiment with the program on a very small scale at this stage before progressing to the next stage, *expansion*. Expansion involves broadening the knowledge management initiative across the organization by increasing visibility to leadership, offices, projects, and disciplines, implementing tools to support different initiatives, and executing strategies to overcome encountered barriers. During this stage, multiple business units, projects, and offices need to start to use the knowledge management processes. Important questions to answer during this stage are how companies can encourage a knowledge-sharing culture among all of its employees and how to support and foster such a culture with incentives. Companies also need to develop strategies of how to educate all employees about the knowledge management initiatives and its processes. Organizations in the *progressive stage* need to work on continuously improving the knowledge management program by measuring performance related to the business goals and objectives. They continue to align incentives with the culture and increase visibility of the benefits of the program. In the final stage, *sustainability*, the knowledge management program should be institutionalized, meaning that it is diffused and used throughout the entire organization, performance is measured, and the knowledge management program is embedded within organizational processes.

Looking ahead to the various stages an organization must go through can help prepare the organization to plan appropriately and strategically for a long-term, viable solution.

Bottom-up implementation

Next to the top down implementation strategy of the STEPS method, organizations also need to support the bottom-up implementation of the knowledge management system from its operational grassroots. The sharing of knowledge is only effective if employees are willing to listen and react to each other. Constructive relations between employees are necessary to enable them to share their insights and freely discuss their concerns. This requires the emergence of micro-communities of likeminded people who are willing to work openly together. Additionally, employees at the operational level will only be inclined to share their knowledge if they see a direct benefit for their day-to-day work processes. The top-down implementation of a knowledge management system, described above, will inevitably fail if employees do not feel comfortable sharing and developing their ideas together. To achieve such an atmosphere of open and effective knowledge sharing, a top-down implementation strategy needs to be accompanied with bottom-up emergent strategies. In the rest of this section, we offer a number of suggestions to companies to support and foster the implementation of knowledge management systems using bottom-up strategies.

First, and most importantly, organizations need to ensure that the benefits of the knowledge management system immediately improve their employees' operations. This recommendation contrasts with the top-down management practice of most of today's implementation efforts that promise a large strategic benefit for the whole company. While it is important to have such a high level vision, it is also critically important to assess the potential benefits of a knowledge management system realistically at the operational level. In this way, employees at the operational level will perceive that the system offers them benefits to improve their work processes. If employees see a direct benefit to their jobs, they will be inclined to start using the knowledge management system to support their ongoing operations and to actively engage in knowledge sharing with peers.

Additionally, organizations need to be realistic when they assess the knowledge that can be exchanged in a knowledge management program and understand the expertise required to train their employees to externalize and internalize knowledge. Only a realistic assessment of what is needed to implement a knowledge management program at the operational level can ensure that employees at the operational level feel at ease with the knowledge management system and, in turn, use it as intended. To enable employees at the operational level to understand the value of a knowledge management program and learn how to play a role in the program, executive management needs to make time available for employees to learn the system, to evaluate it in a practical context, and to discuss the knowledge management program implementation formally and informally with peers.

Because engineering and construction firms operate in a project-based industry, companies need to account for the specific needs of project teams. Project teams usually work in complex, frequently changing environments and knowledge management systems that companies design at the central executive level will not reflect the needs of these complex, project-specific contexts. This also means that companies cannot solely rely on standardized implementation guidelines or training programs developed for non-project contexts to educate their project based work force. In the same line, project teams that implement a knowledge management system should focus on easily reachable goals first, and only slowly improve the sophistication of the system implementation on the project level.

Next to these rather technical ways to support the bottom-up implementation of knowledge management systems, companies need to closely evaluate and, if necessary, change their culture. Effective knowledge management and sharing is not possible in companies with a culture that stresses individual competiveness. Knowledge management requires mutual trust between employees, active empathy for each other, and the willingness among employees to help each other. Companies that assess their internal culture and realize that it is based on individual competiveness need to implement incentive mechanisms that allow for the bottom up emergence of a culture that enables knowledge sharing. For example, companies can revise performance reports and appraisals to include team behavior and willingness to help. Additionally, companies can support and foster mentoring systems, on-the-job training programs, and emotional intelligence training classes. Finally, and maybe most importantly, companies need to make sure that employees realize that they are mutually dependent on each other for success and survival.

Implementation barriers

Part of strategic planning for a successful knowledge management program involves identifying the barriers that the organization may face. The most frequent barriers companies encounter are the lack of time (Carrillo and Chinowsky 2006; Carrillo *et al.* 2004; Fong and Chu 2006) and monetary resources at all organizational levels (Carrillo and Chinowsky 2006; Carrillo *et al.* 2004). Initiating, using and maintaining a knowledge management strategy throughout the organization requires a significant investment of time and money. If organizations are not prepared for this investment or lose their focus during the implementation, the knowledge management strategy will often fail to achieve its original goals.

Another barrier is the lack of support from management and employees (Carrillo and Chinowsky 2006). This includes a lack of unified vision of knowledge sharing and a lack of proactive management strategies (Fong and Chu 2006). Usually cultural differences among employees exist with regard to knowledge sharing. For example, younger generations who grew up with online social networks and wikis might be more inclined to share than their older peers (Chinowsky and Carrillo 2007). These differences, however, do not only exist across generations, but also across cultures.

Poor processes and infrastructure also contribute to knowledge management problems (Carrillo and Chinowsky 2006; Carrillo *et al.* 2004; Fong and Chu 2006). Although many studies cite people, not technology, as instrumental in the knowledge management process, information technology is often essential for employees to store, access, and share knowledge. If the infrastructure is not set up properly, the knowledge management initiative will often fail to meet its intended goals.

Within the AEC sector, the focus on projects can also be a barrier(Chinowsky and Carrillo 2007). Whereas knowledge management initiatives are intended to be organization-wide, the emphasis on project delivery and performance within the industry is often counterproductive. For a knowledge management initiative to gain wide acceptance and use, organizations often must alter their perspective, emphasizing the performance of the organization and the necessity of cross-project learning.

Finally, many knowledge management strategies focus solely on sharing existing knowledge. However, to remain viable, they must also integrate learning to assimilate new knowledge. Chinowsky and Carrillo (2007) indicate that a learning initiative is an integral part of a knowledge management strategy if the strategy is to be viable and sustainable.

Strategies to overcome the barriers

Firms can attempt to mitigate barriers for knowledge management programs. To begin, the commitment and support of top management are critical (Fong and Chu 2006). This often requires a corporate mandate (Chinowsky and Carrillo 2007) to capture the attention of employees and encourage proactive use of the system and an executive management level sponsor for the program. Because one of the most frequently identified barriers is time and money, the organization must continue its commitment for the knowledge management program by dedicating sufficient resources at the start to maintain a viable solution.

In addition, in line with the bottom-up implementation recommendations, it is critical for a successful knowledge management initiative to understand the benefits of knowledge management (Chinowsky and Carrillo 2007; Fong and Chu 2006) and to demonstrate these benefits frequently to management and employees. Obtaining "buy-in" from the organization requires an effective communications strategy (Chinowsky and Carrillo 2007; Javernick-Will 2010). Organizations need to communicate the goals, objectives, vision and features of the selected knowledge management strategy. Firms must also consider organizational incentives to encourage knowledge sharing amongst employees. While there is considerable doubt as to the use of direct incentives for knowledge sharing, if employees are rewarded solely based upon individual or project success versus the success of the overall organization, they will be less inclined to share their knowledge with other peers working on different projects, instead choosing to dedicate their time for their own personal gain.

Conclusion

Since we entered the information age companies must manage knowledge as one of their major assets to maintain their competitive advantage within the global construction market. Before companies can strategically implement a knowledge management system they need to understand what knowledge is, and identify the characteristics of important knowledge to the organization. Some of the characteristics to consider include:

- the explicit or tacit character of the knowledge;
- the complexity of the knowledge;
- how observable the knowledge is;
- how specific the knowledge is to certain contexts; and
- how quickly the knowledge becomes outdated (half-life).

Among these categories, the tacit or explicit character of knowledge is very important because it helps companies to determine how to enable knowledge exchange among their employees. According to the tacit or explicit character of the knowledge, companies can use three knowledge exchange mechanisms:

- combination to exchange explicit knowledge directly;
- externalization and internalization to make tacit knowledge explicit, transfer it through combination, and then convert it back to tacit knowledge; and
- socialization to transfer tacit knowledge directly.

Recent advances in social networking technologies have allowed companies to combine the three transfer mechanisms by designing interactive online platforms where employees can search for and share explicit knowledge, while allowing direct interaction with peers through forums and people searches.

In addition to designing a knowledge management strategy that is custom tailored to the characteristics of knowledge that are important for a company, companies also need to think about how to implement the knowledge management strategy throughout their organization. It is important that companies maintain both a top-down and bottom-up focus during the implementation to be successful. A top-down implementation using the STEPS model is important to align all employees across the company and to establish a global knowledge sharing culture. At the same time, a bottom-up focus is needed to ensure that the knowledge management strategy is beneficial to ensure buy-in for the strategy amongst all employees, particularly those that work at a company's operational level. Once a knowledge management strategy is implemented, companies should ensure continuous quality improvement through ongoing performance measurement and benchmarking activities.

References

Ball, D., McCulloch Jr., W.H., Geringer, J.M., Minor, M.S., and NcNett, J.M. (2006), *International Business: The Challenge of Global Competition*, Boston, MA: McGraw-Hill/Irwin.

Borgatti, S.P. and Cross, R. (2003), "A relational view of information seeking and learning in social networks," *Management Science*, 49(4): 432–445.

Brown, J.S. and Duguid, P. (2001), "Knowledge and organization: A social-practice perspective," *Organization Science*, 12(2): 198–213.

Carrillo, P. and Chinowsky, P. (2006), "Exploiting knowledge management: the engineering and construction perspective," *Journal of Management in Engineering*, 22(1): 2–10.

Carrillo, P., Robinson, H., Al-Ghassani, A., and Anumba, C. (2004), "Knowledge management in UK construction: Strategies, resources and barriers," *Project Management Journal*, 35(1): 46–56.

Chinowsky, P. and Carrillo, P. (2007), "Knowledge management to learning organization connection," *Journal of Management in Engineering*, 23(3): 122–130.

Cook, S.D.N. and Brown, J.S. (1999), "Bridging epistemologies: The generative dance between organizational knowledge and organizational knowing," *Organization Science*, 10(4): 381–400.

Fong, P.S. and Chu, L. (2006), "Exploratory study of knowledge sharing in contracting companies: A sociotechnical perspective," *Journal of Construction Engineering and Management*, 132(9): 928–939.

Garvin, D.A. (1993), "Building a learning organization," *Harvard Business Review*, 71(4): 78–91.

Haindl, G. (2002), "Tacit knowledge in the process of innovation," *Ekonomický casopis*, 01/2002: 107–120.

Javernick-Will, A.N. (2010), "The embedment of a knowledge management program in an engineering organization," *Construction Research Conference 2010 Proceedings*, Canada, online, available at: http://scitation.aip.org/getabs/servlet/GetabsServlet?prog=normal&id= ASCECP0003730411090 00070000001&idtype=cvips&gifs=yes&ref=no.

—— and Levitt, R.E. (2009), "Knowledge as a contingency variable for organizing knowledge management solutions," *2009 ASCE LEAD Conference*, P. Chinowsky and R. Levitt, eds., Lake Tahoe, CA.

—— and —— (2010), "Mobilizing Institutional Knowledge for International Projects," *Journal of Construction Engineering and Management*, 136(4): 430-441.

—— and Scott, W.R. (2010), "Who Needs to Know What? Institutional Knowledge and International Projects," *Journal of Construction Engineering and Management*, 136(5): 546–557.

Kululanga, G.K., McCaffer, R., Price, A.D.F., and Edum-Fotwe, F. (1999), "Learning Mechanisms Employed by Construction Contractors," *Journal of Construction Engineering and Management*, 125(4): 215–223.

Monteverde, K. and Teece, D.J. (1982), "Supplier Switching Costs and Vertical Integration in the Automobile Industry," *The Bell Journal of Economics*, 13(1): 206–213.

Nissen, M.E. (2006), *Harnessing Knowledge Dynamics: Principled Organizational Knowing and Learning*, Hershey, PA: IRM Press.

Nonaka, I. (1994), "A dynamic theory of organizational knowledge creation," *Organization Science*, 5(1): 14–37.

—— and Takeuchi, H.A. (1995), *The Knowledge-Creating Company: How Japanese Companies Create the Dynamics of Innovation*, New York: Oxford University Press.

Polanyi, M. (1967), *The Tacit Dimension*, New York: Doubleday.

Robinson, H.S., Carrillo, P.M., Anumba, C.J., and Al-Ghassani, A.M. (2005), "Knowledge management practices in large construction organizations," *Engineering Construction and Architectural Management*, 12(5): 431–445.

Rockart, J.F. and Short, J.E. (1989), "IT in the 1990s: Managing organizational interdependence," *Sloan Management Review*, 30(2): 7–17.

Spender, J.C. (1996), "Making knowledge the basis of a dynamic theory of the firm," *Strategic Management Journal*, 17(winter): 45–62.

4 A network and culture perspective on organization management

Paul S. Chinowsky

From team to network management

The concept of team is integral to the success of an engineering or construction organization. The idea that a single individual can perform the role of a "Master Builder" in today's complex operating environment is one that is being replaced with a recognition that project success relies on multiple participants working together to achieve a common goal. In this integrated model, the ability for each participant to integrate individual goals and objectives within the overall success criteria is essential for project success. However, integration of goals may not be sufficient to achieve the level of success desired by the project participants or the project owner. In many of today's projects, in both traditional facilities and new project types, issues such as environmental sustainability, long-term flexibility, or international partnerships require a new perspective on project solutions. The ability to develop innovative solutions for these projects that exceeds traditional benchmarks is not only necessary, but is essential to the long-term viability of the project participants. Unfortunately, developing innovative solutions that exceed traditional boundaries is not always easy or at the top of the participant agendas.

One aspect of teams that can be improved to achieve greater levels of success and develop enhanced solutions is to emphasize the relationships within the team construct. Specifically, research is increasingly demonstrating that performance and quality of output is linked to social aspects within a team including trust, reliance, and communication levels. These social aspects are the foundation of the relationships that are developed between team participants. As introduced in this chapter, relationships develop into a network perspective of project and organization development which in turn will be the core of high performance teams that have the ability to produce solutions that exceed traditional measurements.

Culture is overlooked

Facilitating relationship building is a key step in moving toward developing effective organization networks. However, relationships require a foundation on which to develop. A key element of this foundation is a shared culture within an organization. This culture is made up of *the values, beliefs, underlying assumptions, attitudes, and behaviors shared by the people within the organization* (Heathfield 2010). Additionally, the culture is made up of the behaviors that result when the organization arrives at a set of generally unspoken and unwritten rules for working together. Culture is not considered good or bad by itself, but it has a strong influence on the behaviors that individuals take within the organization. The culture can be considered the corporate norms that guide the actions of the individuals within teams and within the overall organization. This influence elevates culture to the important role it plays in guiding teams to perform as a high performance entity.

Culture has been studied in many contexts and in many professional settings. From these studies, the essential elements of culture have been enumerated to assist organization in understanding and developing their own culture (Thompson and Luthans 1990). At the core of these findings are the following essential characteristics:

- Culture is behavior: The interconnection between culture and organizational behavior is strong enough to consider them one and the same in many cases. Specifically, the cultural norms guide organization behavior, and the behavior of individuals within the organization define the organization culture.
- Culture must be learned: Individuals entering an organization at any level must learn the culture of that organization. The time required to learn the culture will vary across organizations and across individuals. However, a guiding principle that should be considered is that the stronger the culture, the longer it will take for an individual to get acclimatized to the culture. However, once the individual learns the culture, the strength of the norms is likely to have a corresponding impact on their behaviors.
- Interaction builds culture: A strong component of transferring culture to existing and new organization members is the interaction that occurs between members on a daily basis. The importance of this interaction requires organization leaders to provide opportunities that allow individuals to interact on a regular basis. In turn, this interaction will result in a transfer of norms that reinforces the culture.
- Subcultures form from performance: In any culture, subgroups form based on individual projects and responsibilities. These subgroups in turn develop subcultures that influence the performance of the subgroup. In a high performance organization, these subcultures can promote and reinforce the desire to perform at a high performance level.

Within the engineering and construction domain, culture has traditionally focused around individual project performance with extraordinary performance often translating to cultural lore that guides the development of organization norms. However, the development of culture at an organization level is often overlooked. As found by Chinowsky and Toole (2009), the lack of organization culture is an issue that organizations are aware of, but find it challenging to alter the current status. Specifically, engineering and construction organizations self-identify that they only moderately agree that they have a culture in place that guides norms, they believe the culture tends to inhibit risk taking, and the organizations lack a culture that encourages interaction and innovation. This contrast between existing culture and desired attributes presents a key challenge to the industry as it strives to introduce high performance teams.

Networks are key to building a high performance organization

The increased demand to move away from traditional levels of performance is driving the need to introduce high performance teams into the engineering-construction work environment. High performance teams achieve outcomes that exceed the expectations of the project and often demonstrate unique or innovative approaches within a final solution. These teams challenge conventional expectations by combining individual strengths and knowledge to generate solutions that exceed the capability of an individual team member. These high performance teams focus on exceeding traditional measures rather than focusing on meeting the benchmark accepted by previous project teams. At the core of these teams is a high degree of connectivity between the team members (Losada 1999). The teams are often characterized by an atmosphere of positive collaboration with significant appreciation and encouragement to other members in the team. The teams create an environment that opens possibilities and encourages action and creativity. These teams are often observed to accomplish complex tasks with ease and grace, whereas low performance teams struggle with their tasks because of negative environments that lack mutual support and enthusiasm. The atmosphere in these teams is often charged with distrust and cynicism.

This concept of high performance is documented and routinely implemented in diverse industries including healthcare and transportation (Poulton and West 1993). However, high performance teams and solutions receive less attention in the construction domain. Rather, the measurement of success within a construction project is often based on meeting historical benchmarks for the classic factors of time, cost, and quality. Changing this perspective is a challenge since high performance teams are characterized by their ability to subsume individual goals in favor of team success and to establish strong elements of trust within the team. These elements are often missing within a construction team, not because of lack of intent, but as described later, because of contractual and industry barriers. The challenge then for the project manager or department manager is to move from the traditional concept of a team to a new construct that builds these underlying connections regardless of the accepted barriers.

The answer to this challenge is to change the team perspective from a group of participants focusing on a project to an integrated group of participants within a network. Within a network, each member follows a set of underlying principles that guides the transfer of information, responsibilities, and outcomes through network interconnections. The result of this guidance is the establishment of a cohesive network where members focus on building long-term relationships that are transferred from activity to activity. Unfortunately, in an area such as construction where independence rather than interdependence is the normal operating procedure, the network will tend to have less cohesion as the associated parties focus on individual plans and goals in equal or greater proportion to the overall network success. However, it is a move to a network perspective where integration is the dominant force that will be the basis of success in the foreseeable future. Developing and understanding this network perspective is the focus of this chapter.

Engineering project organization networks

Networks are found in every aspect of professional activities. Defined as a social network, these networks are social structures composed of nodes that represent individuals or organizations. The nodes are the actors of the structure that are joined by any type of relationship such as "communicates with," "works with," or "trusts." Every actor in the network has the possibility of being linked to every other actor in the network, although this is rare. The degree to which the potential links actually exist in the network is the relative density of the network. The greater the number of actual links that are present, the higher the density level within the network. These networks are in turn assumed to be embedded in a professional or social system in such a way that the system shapes the behavior of the actors as defined by the relationships (Gretzel 2001). Research in a number of fields has shown that social networks operate on many levels, from families up to the level of nations, and play a critical role in determining the way problems are solved, organizations are run, and the degree to which individuals succeed in achieving their goals (Wasserman and Faust 1994).

Understanding how to successfully manage complex networks within a project or firm is a key to moving from a team concept to a network perspective. At the core of this understanding is the ability to model and visualize the actors and their relationships that exist within the project structure. The process of creating and analyzing these networks is referred to as Social Network Analysis (SNA). The fundamental concept behind SNA is the ability to capture the project or organization members and their relationships within a visual map. Using this visualization, the organization can determine strengths and weaknesses based on the level of connectivity, the distances between members, and the relative position of each member within the network structure. Through the mapping and visualization, a network communication model can provide insights into the relationships and interdependencies that form the foundation of high performance teams. Through these insights, an organization can focus upon the development of the critical associations that may be missing which in turn are inhibiting the project or firm from operating as a high-performance team.

This approach to understanding and building social networks has been proven to be helpful in understanding and improving organizations in many diverse areas. Classic network research focused on sociological networks involving individuals in the workplace and their exchange of information to complete tasks (Krebs 2004). Additional studies focused on international relationships in areas such as research collaboration and international investment (Krebs 2004). Construction engineering and management researchers have utilized network analysis to examine issues such as the emergence of cultural boundary spanners in global engineering services networks (Di Marco *et al.* 2009) and the structure and relationships within project organizations (Chinowsky *et al.* 2008). Within these studies, the ability to map participant relationships within a structure that can be visualized using network analysis software is a significant benefit to organization leaders.

In the project management domain, the use of social network analysis has emphasized project communications and the role communications can play in assisting coordination functions (Pryke and Smyth 2006). However, communication is only one factor that can be modeled with SNA tools and methods (Morton *et al.* 2006; Katsanis 2006). As outlined in the Social Network Model for Construction, human dynamics factors including reliance, trust, and values augment traditional communication analysis when elevating the project analysis to include knowledge sharing and high performance outcomes (Chinowsky *et al.* 2008). Additionally, the understanding that engineering projects are fundamentally unstable networks that get reinitiated for each project is changing the focus on what constitutes a successful network team. Understanding how to balance this instability with the need to implement high-performance team concepts is the challenge for organization leaders. However, in contrast to traditional management approaches in the construction industry that advocate optimizing project structures for individual projects, the network approach introduced here emphasizes understanding the larger need to integrate an organization for the benefit of long-term performance and long-term collaboration.

Network measures that impact high performance

Graph theory, or the study of graphs, is the mathematical foundation on which SNA is developed. In its simplest terms, graph theory consists of a body of mathematical formulas that describe the properties of the patterns formed by lines connecting points (Scott 1991). It uses matrix principles and operations to calculate network properties and characteristics through pair-wise relations between objects from a certain collection. SNA is a powerful tool because it allows users to visually analyze network properties and characteristics as opposed to analyzing mathematical matrices (Hanneman and Riddle 2005).

The use of graph theory as a basis for SNA introduces a wide range of potential measurements that can be used for an individual analysis of an organization or project graph. However, only a partial set of these is commonly used by engineering and construction researchers when examining the characteristics of a project or organization network. This is primarily due to the need to focus on fundamental issues such as the amount of a relationship that exists in a network, the relative position of a person in a network, or the influence of individuals within a network. The reduced set of measurements commonly used in this form of analysis is as follows:

Network density – A measure to indicate the amount of interaction that exists between the network members. Density reflects the number of actual links that exist between members in comparison to the number of potential links that exist if all members were connected through relationship links. The larger the density number that is calculated, the greater the number of relationships that actually exist in the network. This measure can vary from 0 to 1, with 1 indicating a complete graph, or one in which all points are connected with one another. The authors have found that a density of around 25–50 percent indicates a greater likelihood for a high performance team to exist within the network.

Centrality – A measure that reflects the distribution of relationships through the network. Centrality counts the number of incoming and outgoing links that exist for a given actor in the network. In a highly centralized network, a small percentage of the members will have a high percentage of relationships with other members in the network. In contrast, a network with low centrality will have relatively equal distribution of relationships through the network. The "kite network" shown in Figure 4.1 is used to explain centrality (Krackhardt 1990). In this network, two nodes are connected if the individuals speak weekly. For example, A speaks with C on a weekly basis, but not with I. Therefore, A and C are connected by lines, but there is no link drawn between A and I.

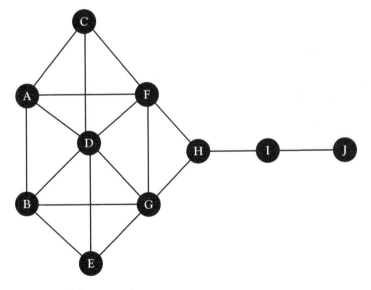

Figure 4.1 A kite network.

In the kite network, D has the most direct connections in the network, making this actor the most active in the network. This actor is a connector or hub in this network, has a high centrality, and therefore is central to the network. Using SNA software, it is possible to view the network via sorting by degree. When this is done, the nodes with the higher degrees are enlarged while nodes that are less central are made smaller, reflecting their respective values of degree centrality. This allows observers to quickly distinguish which actors are "in the thick of things" and which ones are more peripheral to the network. Taking this concept and applying it to the kite network yields the network shown in Figure 4.2.

- **Power** – The power variable works in conjunction with centrality. Whereas centrality measures the total number of relationships that an individual has in the network, power reflects the influence of an individual in the network. Individuals who are giving information to others in the network, who are in turn passing along that information to others, have a high degree of influence or power. Individuals who are mainly on the receiving end of communications may be central in the network, but have little power as they do not influence the actions taken by others.
- **Betweenness** – This variable measures the amount of information that is routed through an individual to distribute to the team. Betweenness measures how often a given actor sits "between" others, with "between" referring to the shortest network path. Betweenness counts the number of paths that pass through a node. Interactions between two nonadjacent actors can depend heavily upon the actors who lie on the path between the two. An actor that is on many paths is assumed to have a higher likelihood of being able to control information flow over the network. This rating indicates which individuals are involved in discussions that are occurring within the network.

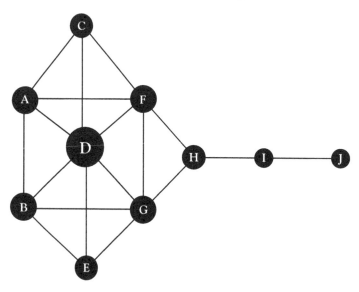

Figure 4.2 Kite network sorted by degree centrality.

Networks within organizations

Before an effective organization network can be developed and in turn high-performance achieved, it is essential to understand that organizations are comprised of multiple networks. Some of these networks are obvious, such as which members communicate with each other on a regular basis, or which members are located in a similar geographic region, or which members are in a similar discipline. Other networks are obvious, but may not necessarily be ones that have been considered as important such as which members have experience working with each other on a professional basis. In contrast, many networks are less obvious yet have an equal or greater impact on the organization and its performance. These networks include information-based networks such as which members exchange information on a project, social-based networks such as which members share values, and project-based networks such as which members need to communicate on a task. Each of these types of networks exists in every organization and is equally important in developing high performance. The examples used in this section are based on an engineering firm with six offices and an extensive history. The organization has grown steadily and has received numerous engineering awards for its work.

Information-based networks

Information-based networks focus on the information and knowledge that is exchanged in the context of developing project solutions or enhancing organization performance. These networks are the measurable characteristics that affect project efficiency. Examples of these information-based networks include:

- **Communication networks** – Communication in this context is the most general of the information networks. Communications are measured to determine the informal network that exists within a project or organization team. Informal networks are critical due to their ability to activate when unexpected problems arise (Katzenbach and Smith 1993). Therefore, the first step for the organization is to establish a social network that has communications connections that extend beyond the formal hierarchy and connect as many personnel as possible. Figure 4.3 illustrates the communication network for the organization.
- **Information networks** – Information networks include the members that an individual member interacts with to complete specific tasks. In these networks, information is exchanged in two directions. In one direction, a member has a key set of individuals from whom information is obtained to assist in completing assigned tasks. In the opposite direction a member provides specific information to others to assist them in completing their required tasks. These networks may be different depending on the tasks, levels of experience, and the impact of the social dynamics. These networks are important because they are an indicator of the efficiency of information transfer within an organization. If individuals are obtaining and distributing information over a wide network, then information transfer becomes more efficient as bottlenecks are reduced and the informal network begins to operate. Figure 4.4 illustrates the information network for the organization indicating a drop in density from the general communication network.
- **Knowledge networks** – Knowledge networks are the strategic component for achieving high performance results. To move from a reactive project process to a proactive process,

teams must transfer the exchange focus from information to knowledge. In this transfer, the team moves its focus away from simple task implementation and individual goals to how the organization and tasks can be improved for mutual benefit (Chinowsky and Carrillo 2007). This level of interaction is difficult to achieve until the concurrent social networks, trust and value sharing, are achieved within the organization. Figure 4.5 illustrates the knowledge network indicating a significant reduction in density from the previous network reflecting the difficulty in developing knowledge exchange in organizations, even ones with long operating histories.

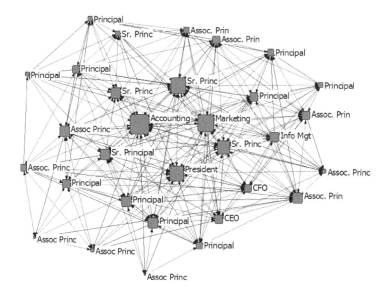

Figure 4.3 Informal communication network within the sample organization.

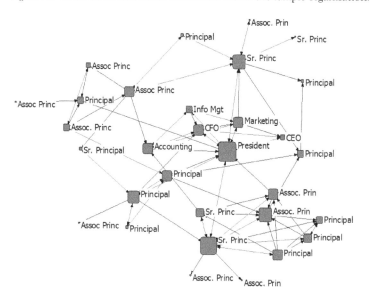

Figure 4.4 Specific communication network within the sample organization focusing on organization specific information transfer.

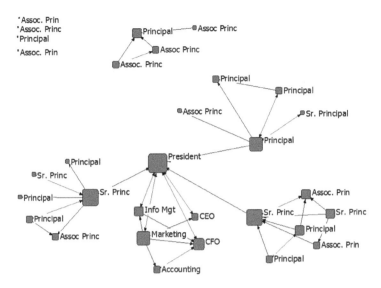

Figure 4.5 Knowledge exchange network within the sample organization focusing on organization specific knowledge transfer.

Social-based networks

The second of the network types, social-based networks, focuses on the motivators for individuals to increase performance in an organization or on a project. The need to identify these networks is based on research findings that high performance teams require trust and shared values to achieve the knowledge sharing which results in enhanced solutions (Kotter 1996). However, the density within these networks is often more difficult to increase than information-based networks as factors such as project instability and geographic distribution often hinders the development of social-based relationships. Therefore, social-based networks include several layers of relationships through which an organization often progresses while striving for the preferred goal of shared values.

- **Experience networks** – The familiarity that exists between individuals through previous work experiences affects the level of interaction and trust that individuals develop within networks. People who are new to an organization network generally do not feel comfortable exchanging anything beyond required information until they are fully assimilated and have developed a favorable perception of other network actors (Cross *et al.* 2002). Given this obstacle, ensuring that a network has a mix of individuals who are new and have previous experience is critical to "seeding" a successful organization or project initiative.
- **Reliance networks** – A project schedule is a blueprint for reliance on any type of project. In this blueprint, dependencies detailed on a schedule dictate what information is to be exchanged within the project network and when this exchange is going to take place in the context of the overall project. These dependencies establish reliance between network members in that one member is reliant on another to complete their task and provide required information for the next member to complete their required

task. A fundamental requirement in such a project network is that each member believes that they can rely on other members to complete their given tasks based on their skills and competence (Blois 1999). Reliance networks map the perceived reliance dependencies that exist within an organization. These mappings assist in determining if an individual is overly central in the organization or if individuals may be relying on unexpected sources of information. Figure 4.6 illustrates the reliance network from the engineering firm example.

- **Trust networks** – The third level of social networks is the concept of trust networks. Trust and reliance are often confused and are frequently used interchangeably. Although there are multiple differences between the two concepts, the relevant difference in the social networks is expectation. Specifically, in a trust relationship, one member of the network trusts that another member will go beyond just completing their task to act in a manner that is mutually beneficial to both parties. This trust leads to an emotional connection within the network whereby if one member does not act in a mutually beneficial manner, then the other member feels let down. The importance of this trust concept is that a member who trusts another individual to work for mutual benefit will have a greater likelihood of sharing knowledge than an individual who believes that no mutual benefit will occur. Figure 4.7 illustrates the trust network from the engineering firm indicating a strong level of trust between the network members.

- **Value networks** – The goal within any organization striving to achieve high performance is to have members share values. These values include both social values such as responsibility, integrity, honesty, morality, quality, and timeliness, as well as context values such as client interaction, worker treatment, and environmental stewardship. If a network can be formed where the members share the contextual values as well as a segment of their social values, then they will reach the performance context that is required to fully share knowledge and achieve high performance results (Katzenbach and Smith 1993).

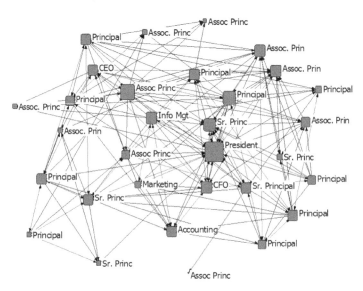

Figure 4.6 Reliance network within the sample organization focusing on which individuals rely on others to get their individual tasks completed.

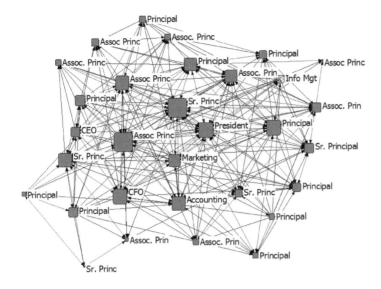

Figure 4.7 Trust network within the sample organization focusing on which individuals trust others to use knowledge for mutual benefit.

Networks across organizations

Networks are not confined to relationships within an organization. Equally important relationships can be mapped and analyzed across organizations where similar social and information relationships exist as well as additional issues such as coordination, learning, and cultural differences. In these networks, the focus transfers to facilitating communication and relationships between organizations to achieve high performance in interorganization relationships.

- **Project networks** – Whereas information and social networks describe relationships within an organization or project based on general interactions, project networks describe specific relationships within a project context. The focus of these network mappings is to determine if individuals are communicating in accordance with required tasks. Two perspectives are essential in examining project networks. First, the schedule-based perspective must be considered where the focus is on executing specified tasks within the project schedule. Second, the innovation perspective focuses on the opportunities a project team has to exchange knowledge in an effort to innovate and learn through project activities. Finally, it should be noted that a contractual perspective must be considered to determine where legal constraints limit the ability of teams to freely interact and exchange knowledge.

  The first of these perspectives, the schedule perspective, emphasizes the role of the schedule to define network links. The last two decades have been witness to a significant amount of research on project task network scheduling using computerized systems. Research in these areas has been augmented with information processing capacity to enable the application of fuzzy logic (Ayyub and Haldar 1984; Lorterapong and Moselhi 1996), integration with geographic information systems (Poku and Arditi 2006) and three-dimensional computer aided design models (McKinney and Fischer 1998), and Monte Carlo simulation approaches (Lee 2005) in task network analysis research. However, this task-centric approach to the management of projects neglects the important interface management function of project management (Morris 1994). As the number of tasks in the project increase and the number of project participants increase, the need to coordinate these parties to enable tasks to be completed becomes increasingly important and difficult. The schedule perspective on networks emphasizes whether the appropriate task-organization alignment is in place to enable efficient exchanges and, hence, effective project execution (Chinowsky and Taylor 2009). In this perspective, task dependencies in a network are mapped to determine which tasks are dependent upon others and which individuals are responsible for the dependent tasks (Figure 4.8). Using these relationships, the appropriate level of communication and knowledge exchange between these individuals can be determined and then compared against what is actually occurring in the project. This will enable the project manager to determine which tasks are vulnerable to coordination misalignment.

The second of the project perspectives focuses on a key element of developing inter-organization high performance, the ability of project teams to exchange knowledge which leads to learning and innovation. Firms must learn to become effective at executing new processes or using new technologies. However, some of these processes span organizational boundaries and require firms to adapt together. These systemic innovations necessitate multiple firms to change their practices concurrently (Taylor and Levitt 2004). Although these innovations can enable significant increases in overall productivity, they are difficult to implement when relationships between firms in a project network may be unknown, unidentified, or unstable across multiple projects (Taylor and Levitt 2007). For this reason, the mapping of relationships between firms on a project can highlight the connections where opportunities for learning exist and where connections need to be enhanced to facilitate greater learning opportunities.

As noted, interorganization project networks have an inherent barrier that must be considered in the construction industry, contractual relationships. As well known and often discussed in the industry, contracts are not only definers of project scope and limitations, but can also serve as barriers to team interaction. Specifically, while the exchange of knowledge leads to effective projects, the exchange of knowledge can be hindered by contractual clauses which introduce liability when such information is exchanged. While there is no current solution for this issue, it is one that should be at the forefront for any organization that strives to enhance its operations through inter-organizational learning.

• **Culture networks** – Weak interpersonal relationships in culturally diverse teams will impede adequate knowledge exchange processes within teams (Luo 2001). The focus of culture networks is the identification of project links across organizations that include members originating from different cultures. In this age of globalization and international teams and projects, the identification of these multicultural networks can be significant in developing successful teams. Of particular interest in cross-cultural collaborative engagements is the identification of another form of team participant other than just the team leader (Ansett 2005). Luo (2001) concluded that because members with varying cultural backgrounds find it difficult to communicate within cross-cultural joint ventures, having a number of culturally related group members allows for seamlessness in coordination and communication. To adapt to the globalizing cross-cultural team working environment, many firms have chosen to designate some team members to

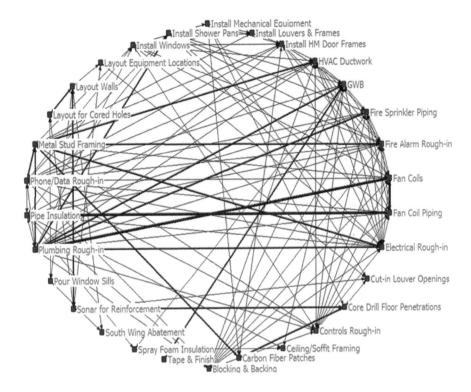

Figure 4.8 Project task network combining responsibility for tasks with communication requirements between project stakeholders.

bridge the gap between team members with different backgrounds (Cross and Parker 2004). Adopting this practice is in line with the suggestion that organizations should appoint certain persons to span boundaries between units (Aldrich and Herker 1977; Friedman and Podolny 1992).

The importance of this designation originates from the exploration of cultural boundary spanner emergence which finds that differences in national cultural backgrounds give rise to national cultural boundaries (Di Marco *et al.* 2009). These boundaries result in knowledge system conflicts that are detrimental to collaboration performance. Recent research on global collaborations also supports this argument (Bryant 2006; Mahalingam and Levitt 2007). We find that when potentially detrimental knowledge system conflicts occur, nominated cultural boundary spanners can mitigate them by negotiating boundaries and thus forming new "joint fields" that enable team members from different national cultural backgrounds to pursue common goals (Levina and Vaast 2005: 337).

To formally define this role, a cultural boundary spanner in cross-cultural project networks is a member of the project team that provides vital cultural insight and background that the entire network draws on to get its work done. Additionally, cultural boundary spanners can be any team member who connects the members of culturally distinct subteams in project networks through their knowledge of the collaborative counterparts' backgrounds. The identification of these individuals, in conjunction with the identification of project network relationships, will enhance the likelihood that cross-cultural projects will continue according to the identified schedule.

Although the importance of cultural boundary spanners in multinational teams can be apparent, one should not dismiss the role of such individuals in domestic project teams. Specifically, a cultural boundary spanner can also be an individual who works to cross the divide between organizations with significantly different cultures. For example, a large general contractor may have a very different culture from a specialty subcontractor working on a project. An individual with experience in both sides of the arrangement can provide significant benefits by bridging the cultural divide between the organizations and reducing the time required for each to learn the other's processes and preferences.

In each context, the role of the cultural boundary spanner should not be minimized. This new area of analysis is clearly identifying such individuals as critically important in both international and domestic contexts. Organizations that can identify such individuals prior to the start of new efforts will have a distinct advantage in setting the stage for a successful project.

Networks across scale

The array of networks that exist within an organization, between organizations, and at multiple levels within an organization can challenge leaders to determine the appropriate size and diversity that is proper for any given network. The answer to this concern is not a specific value. Rather, networks can exist at any scale, from small project teams to large inter-organization, global networks. Both social-based and information-based networks can be defined within the context of small or large-scale networks. The definition of a small or large-scale network is arbitrary. Depending on the focus of the analysis, large-scale networks can be 100 members or 1,000 members. In the discussion presented here, the division between small- and large-scale networks is based on the concept of banding. Research in psychology has demonstrated that humans operate effectively in bands of networks with a preferred number of no more than 30 members (Dunbar 1993). This amount is equal to the number of maximum people that an individual can keep in close contact with on a regular basis. Concurrently, most people have a limit of 100 people, roughly three bands, which they can communicate with on a regular, but infrequent basis. Given this banding concept, the threshold for a small-scale network adopted in this discussion is equal to the three-band limit, or 100 members. Once this number is exceeded, it is reasonable to assume that separation will enter the network where individuals are no longer able to stay in contact with the entire network. At this point, the network transfers from a small group dynamics focus to a large-scale organization focus.

Small-scale networks

Small-scale networks primarily focus on either project networks that are limited to supervisory personnel or firm networks focused on either a limited number of divisions or supervisory personnel. Given the definition that a small-scale network requires individuals to have the opportunity to stay in contact with all other members on at least an infrequent basis, the focus of small scale networks is on the establishment and retention of relationship links between the members. From a managerial or leadership perspective, the management of these relationships is dependent on the individuals who are placed in positions of centrality within these networks. Of particular interest are the common small-scale network deficiencies that are illustrated in Figure 4.9.

- **Gatekeeper** – A gatekeeper is an individual who controls the flow of information within the network. In a small-scale network, a gatekeeper can significantly reduce the effectiveness of the network by restricting the flow of information to network members.
- **Fragility** – Network fragility occurs when a single individual is the only link to a group of other individuals within the network. The loss of this individual can seriously impact the ability of the network to retain its connectivity.
- **Distance** – The key to small network effectiveness is the establishment of multiple connections from each individual to others in the network. A key reason for this connectivity is to reduce the number of steps required for one individual to communicate with another. However, when an individual does not establish these links, then the distance increases between individuals in the network. This can result in individuals feeling isolated and reducing their input into project organization solutions.

As illustrated, small-scale networks emphasize connectivity and interaction. In a project context, this means a project network that emphasizes interaction between members and project managers that understand the importance of collective leadership and input from each individual. In an organization context, this translates to an organization that is collaborative and incorporates managers who understand the value of cross-communication and the transfer of solutions across discipline and geographic boundaries.

Large-scale networks

Large-scale networks focus on interorganizational projects or management networks in large organizations. These networks incorporate the same potential issues that exist in small-scale networks, but the number of individuals in the network provide greater robustness due to the inherent likelihood that connections will exist between individuals in different sectors

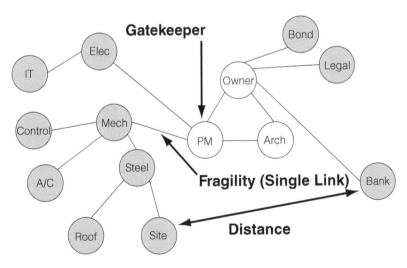

Figure 4.9 Illustration of issues that can arise in a small-scale network.

within the network. From this perspective, the emphasis of the network management alters to emphasize the reduction of cliques within the network. Specifically, cliques are individual groups that have multiple links within that group, but have minimal connection to the overall network. This isolation is very common in organizations where cliques can form from either geographic or discipline origins.

The management of large-scale networks requires individuals to have a collaborative perspective that values input from all areas within the network. The concept of a collaborative perspective is one that differs from classical, hierarchical management theory. Specifically, a collaborative perspective focuses on everybody in the network bringing an equal value to a project or organization concern. Within a global organization, this perspective requires a central office to consider input from remote geographic offices to be as important as the ideas being generated from the central staff. In multi-division organizations, this perspective requires organizations to extend opportunities and seek input from remote divisions, small divisions, and new divisions as well as the traditional central core of the organization.

These examples illustrate a central theme of large-scale networks, each arm or band of the network has equal importance in the overall success of the organization. Unfortunately, this perspective is often lost in a large-scale network where it is easier to focus on the central core of relationships (the home office or primary division) than spending the resources required to include the different elements within the overall network. Responding to this issue and changing the course of the network collaboration takes concentrated effort by the organization leaders. Specifically, in a large network environment, leaders must identify where further connections between areas within the network can be developed and fostered. Once these connections are developed, the leaders must then encourage the use of these connections as fundamental transfer points within the organization. This is the area that is the challenge in this environment. While it is an operational issue to develop relationships, it is a strategic issue to value and utilize these relationships. In this perspective, the development of an effective and long-lasting large-scale network becomes a strategic issue that is central to the long-term success of the overall network.

Implementing a network perspective in engineering project organizations

Once an organization decides to adopt a network perspective, the logical and most often stated next question is how to start moving in that direction. The sections in this chapter provide an overview of the types of networks that exist in an organization and the perspectives that can be adopted when moving to a high performance, network perspective. To emphasize the steps needed to approach this perspective and to provide a succinct overview of the elements required to retain this focus, the following ten steps are an overview for focusing on a network perspective within the organization.

1. **Determine appropriate network members** – Every member in an organization does not have to be a part of every network. However, every member is a potential member of several organization networks. The first step in adopting a network perspective is to determine which networks each member of the organization should be included within.
2. **Focus on relationships** – Networks are based on relationships between members. Whether the relationship is tangible such as communication frequency, or intangible such as level of trust, the relationships are the key to network definition. As such, defining the relationships that are represented by the network is the second element of network development.

3. **How wide is the net?** – Relationships and multiple potential membership options can establish a potential for a broad set of members within an individual network. This may or may not be appropriate depending on the analysis. Determine the appropriate number of members in a network before continuing.

4. **Understand the objective** – What is it that is being analyzed by constructing a network map? Understanding the objective of the network is a foundation for determining how to analyze the network.

5. **Select the measurements** – A wide variety of measurements are available for network analysis. Determine the ones that are appropriate for the analysis that is being undertaken.

6. **Participation is key** – The analysis of any network requires full participation by the individuals within the network. Although an external perspective can be used, participation by network members is always preferable.

7. **Build on the results** – Network measurements provide an insight into how an organization is operating. Building on these results to implement needed changes is the key to continued support for network analysis.

8. **Gatekeepers and fragile networks** – Network analysis indicates areas where high performance is lacking and where the network may be susceptible to repeated poor performance. The elimination of network deficiencies should be a priority.

9. **Return engagements** – A network analysis is one snapshot in time. Regular analyses of the organization are essential to determine if improvements or setbacks are occurring within the network.

10. **Commitment** – Organization networks are complex organisms that take many years to develop. Changing the relationships within a network also take time. Have patience – change will occur, but it takes time.

Keys to sustaining a network culture to achieve high-performance organizations

Once a network culture is developed in an organization and an understanding is in place for measuring the networks, the last thing for an organization leader to consider is how to sustain the high-performance culture. Although every organization has its unique features, several common keys exist to sustaining the high performance culture.

- **Reinforcing culture** – A high-performance culture is an investment in multiple resources over multiple areas within the organization. Time, money, and mentoring are only a few of the resources that are devoted to building high-performance as a core value in the organization. However, the value of this investment will only be realized once the high-performance culture is reinforced throughout the organization to both current and future members. This reinforcement of the culture is a core requirement for sustaining the high-performance organization.

- **Reinforcing trust** – The role of trust in achieving high performance is significant. However, in contrast to many popular opinions, professional trust does not automatically develop out of company picnics or dinners. Rather, professional trust develops out of reliance and dependency on others. Given this reliance and dependence, individuals will develop trust once others demonstrate that the trust is founded on performance. Furthering this concept, it is imperative for the organization to reinforce trust-building scenarios through interdependency at both the division and office levels.

- **Reinforcing collaboration** – Collaboration requires a combination of trust and communication. Each of these elements is a unique network as introduced previously. However, each of these networks presents a challenge as it in turn requires network members to build these elements from both social and technical perspectives. Given these challenges, organization leaders must take an active role in ensuring that the fundamentals of the collaborative relationships are continually reinforced. Whether these relationships span departments, organizations, or cultures, the common thread remains that high performance emerges as collaboration is sustained as a result of successful trust and communication relationships.

References

Aldrich, H. and Herker, D. (1977), "Boundary Spanning Roles and Organization Structure," *The Academy of Management Review*, 2(2): 217–230.

Ansett, S. (2005), "Boundary Spanner: The Gatekeeper of Innovation in Partnership," *Accountability Forum*, 6: 36–44.

Ayyub, B. and Haldar, A. (1984), "Project Scheduling using Fuzzy Set Concepts," *Journal of Construction Engineering and Management*, 110(2): 189–204.

Blois, K. (1999), "Relationships in Business-to-Business Marketing – How is their Value Assessed?" *Marketing Intelligence and Planning*, 17(2): 91–102.

Bryant, P.T. (2006), "Decline of the Engineering Class: Effects of Global Outsourcing of Engineering Services," *ASCE Journal of Leadership and Management in Engineering*, 6(2): 59–71.

Chinowsky, P.S. and Carrillo, P. (2007), "The Knowledge Management to Learning Organization Connection," *Journal of Management in Engineering*, ASCE, 23(3): 122–130.

Chinowsky, P.S. and Taylor, J. (2009), "Task Interdependency and Communication Networks," *Proceedings of the Fourth Specialty Conference on Leadership and Management in Construction*, Lake Tahoe, CA, November, 2009.

Chinowsky, P.S. and Toole, T.M. (2009), *Enhancing Innovation in the EPC Industry*, Research Summary, Construction Industry Institute.

Chinowsky, P.S., Diekmann, J., and Galotti, V. (2008), "The Social Network Model of Construction," *Journal of Construction Engineering and Management*, 134(10): 804–810.

Cross, R.L. and Parker, A. (2004), "The Hidden Power of Social Networks: Understanding How Work Really Gets Done in Organizations," Boston, MA: Harvard Business School Press.

——, ——, and Borgatti, S.P. (2002), "Making Invisible Work Visible: Using Social Network Analysis to Support Strategic Collaboration," *California Management Review*, 44(2): 25–46.

Di Marco, M., Taylor, J., and Alin, P. (2009), "Cultural Boundary Spanning in Global Project Networks," *LEAD Specialty Conference on Global Governance in Project Organizations*, Lake Tahoe, CA.

Dunbar, R.I.M. (1993), "Coevolution of Neocortical Size, Group Size and Language in Humans," *Behavioral and Brain Sciences*, 16 (4): 681–735.

Friedman, R.A. and Podolny, J. (1992), "Differentiation of Boundary Spanning Roles: Labor Negotiations and Implications for Role Conflict," *Administrative Science Quarterly*, 37(1): 28–47.

Gretzel, U. (2001), "Social Network Analysis: Introduction and Resources," online, available at: http://lrs.ed.uiuc.edu/tse-portal/analysis/social-network-analysis/ [accessed July 2010].

Hanneman, R.A. and Riddle, M. (2005), *Introduction to Social Network Methods*, Riverside, CA: University of California, Riverside.

Heathfield, S.M. (2010), *How to Build a Teamwork Culture*, online, available at: http://humanresources.about.com/od/involvementteams/a/team_culture.htm.

Katsanis, C.J. (2006), "Network Organizations: Structural and Strategic Implications," *Proceedings of the 2nd Specialty Conference on Leadership and Management in Construction*, May 2006, Louisville, CO: PM Publishing, pp. 108–115.

Katzenbach, J. and Smith, D.K. (1993), *The Wisdom of Teams*, Boston, MA: Harvard Business School Press.

Kotter, J.P. (1996), *Leading Change*, Boston, MA: Harvard Business School Press.

Krackhardt, D. (1990), "Assessing the Political Landscape: Structure, Cognition, and Power in Organizations," *Administrative Science Quarterly*, 35(2): 342–369.

Krebs, V.E. (2004), "Managing the Connected Organization", online, available at: www.orgnet.com/MCO.html [accessed July 2010].

Lee, D. (2005), "Probability of Project Completion Using Stochastic Project Scheduling Simulation," *Journal of Construction Engineering and Management*, 131(3): 310–318.

Levina, N. and Vaast, E. (2005), "The Emergence of Boundary Spanning Competence in Practice: Implications for Information Systems' Implementation and Use," *MIS Quarterly*, 29(2), 335–363.

Lorterapong, P. and Moselhi, O. (1996), "Project-Network Analysis Using Fuzzy Sets Theory," *Journal of Construction Engineering and Management*, 122(4): 308–318.

Losada, M. (1999), "The Complex Dynamics of High Performance Teams," *Mathematical and Computer Modelling*, 30(9): 179–192.

Luo, Y. (2001), "Antecedents and Consequences of Personal Attachment in Cross-Cultural Cooperative Ventures," *Administrative Science Quarterly*, 46(2): 177–201.

Mahalingam, A. and Levitt, R.E. (2007), "Institutional Theory as a Framework for Analyzing Conflict on Global Projects," *ASCE Journal of Construction Engineering and Management*, 133(7): 517–528.

McKinney, K. and Fischer, M. (1998), "Generating, Evaluating and Visualizing Construction Schedules with 4D-CAD Tools," *Automation in Construction*, 7(6): 433–447.

Morris, P. (1994), *The Management of Projects*, London: Thomas Telford.

Morton, S.C., Dainty, A.R., Burns, N.D., Brookes, N.J., and Backhouse, C.J. (2006), "Managing Relationships to Improve Performance: A Case Study in the Global Aerospace Industry," *International Journal of Production Research*, 44(16): 3227–3241.

Poku, S. and Arditi, D. (2006), "Construction Scheduling and Progress Control Using Geographical Information Systems," *Journal of Computing in Civil Engineering*, 20(50): 351–360.

Poulton, B.C. and West, M.A. (1993), "Effective Multidisciplinary Teamwork in Primary Health Care," *Journal of Advanced Nursing*, 18(6): 918–925.

Pryke, S. and Smyth, H. (2006), *The Management of Complex Projects: A Relationship Approach*, Malden, MA: Blackwell Publishing.

Scott, J. (1991), *Social Network Analysis: A Handbook*, London: Sage.

Taylor, J. and Levitt, R. (2004), "Understanding and Managing Systemic Innovation in Project-based Industries," in Slevin, D., Cleland, D. and Pinto, J. (eds.), *Innovations: Project Management Research*, Newton Square, PA: Project Management Institute, pp. 83–99.

Taylor, J. and Levitt, R. (2007), "Innovation Alignment and Project Network Dynamics: An Integrative Model for Change," *Project Management Journal*, 38(3): 22–35.

Thompson, K. R. and Luthans, F. (1990), "Organizational Culture: A Behavioral Perspective," in B. Schneider (ed.), *Organizational Climate and Culture*, San Francisco, CA: Jossey-Bass pp. 319–344.

Wasserman, S. and Faust, K. (1994), *Social Network Analysis*, Cambridge, MA: Cambridge University Press.

5   Strategic innovation in EPC

T. Michael Toole, Paul S. Chinowsky, and Matthew R. Hallowell

Introduction

Stop and look around you. Most likely you are sitting in a room surrounded by floor, wall and ceiling materials that were not available 80 years ago. Behind these materials are mechanical systems—heating, ventilation and air conditioning (HVAC), plumbing and electrical materials and devices—that are very different from those installed 50 years ago. Now stop and replay the last few hours of your life. Most likely you used a mobile phone, checked email, or surfed the net using a personal computer. If you drove a car, you drove a machine that offered significant performance advantages over cars manufactured 30 years ago and cost less in terms of real dollars.

What is the point here? The point is that almost every minute of the day, every aspect of your life is affected by the innovation that some firm successfully pursued and embodied in something that you bought or came in contact with. One of General Electric's most successful marketing slogans was, "We bring good things to life." If we stop and examine our lives, we realize that everything we do is touched in a big way by a myriad of firms all continuously pursuing innovation to bring "good" things into our lives.

As is discussed in nearly every chapter of this book, today's business environment is changing so fast that firms that do not pursue innovation quickly lose their efficiency, their effectiveness, their customers, their employees and so on. Many researchers and journalists outside the Engineering-Procurement-Construction (EPC) industry point at the industry as one in which very little innovation occurs. Researchers and industry leaders know that innovation does occur in EPC; it is just that the pace of innovation is slow and signs of change may be hard to see for outsiders. What many EPC leaders fail to recognize, however, is that tomorrow's most successful EPC firms will be those who manage innovation better than do their peers. Innovation does not just happen. Indeed, there are many firm- and industry-wide factors that impede innovation. Successful EPC leaders will be those that intentionally and effectively manage their organization's ability to innovate.

This chapter will help EPC managers manage innovation by presenting practical guidance from existing books and articles on innovation, as well as the findings of a three-year research project that was sponsored by the Construction Industry Institute and conducted by a team of seasoned industry veterans and academics (including the authors). The research consisted of a literature search on innovation both within and outside of EPC, two dozen pilot interviews, a survey of over 150 EPC professionals, development of a tool to assess an organization's innovation capabilities, and pilot testing of this tool within five firms.

The chapter starts with brief discussions of why innovation is so important, the types of innovation that occur, and the barriers that hinder innovation in EPC. The bulk of the chapter discusses the organizational characteristics that enable innovation yet are often missing in EPC organizations. A key lesson for successful leaders is that they must be willing to invest resources now to achieve long-term innovation goals even when the outcomes are highly uncertain.

Definitions and types of innovations

Before we discuss why an organization's innovation capability is important, it is appropriate to make sure we have an understanding of what innovation is. As is true for the terms *leadership*, *quality*, *high performing* and many other important business terms, there is not one simple definition of innovation on which everyone agrees. Below are overlapping concepts of innovation from various authors.

- In his classical book, *The Economics of Industrial Innovation*, first published in 1974, Freeman (1982) defined innovation as any improvement in a process, product, or system that is novel to the institution developing the change.
- Park *et al.* (2004: 170) define innovation as "the generation, development, and implementation of ideas that are new to an organization and have practical or commercial benefits."
- Dikmen *et al.* (2005: 81) wrote, "All kinds of improvements within the construction value chain and business system, as a result of new ideas leading to higher value for the client, wealth to stakeholders, or higher competitive advantage, should be considered innovations."
- The European Organisation for Economic Co-operation and Development's definition is "the implementation of a new or significantly improved product (good or service), or process, a new marketing method, or a new organizational method in business practices, workplace organization or external relations" (OECD/Eurostat 2005: 46).

It should be clear from the definitions provided above that innovation need not be technology-based. Changes that provide value to customers and stakeholders can be based on new marketing approaches, organizational structure, progressive human resource policies and programs, and—as Wal-Mart has so aptly demonstrated—through new systems for procurement, distribution, warehousing, and shelf stocking. It should also be apparent from the above definitions that innovation is not about internal research and development. Thirty years ago, the most innovative firms typically had large R&D labs. As Chesbrough (2003) and others have written, new products and services today are often a product of *open innovation*, in which firms collaborate with suppliers, customers, and competitors to generate new products and services. This bodes well for the EPC industry as it has traditionally had one of the lowest percentages of R&D investment among all industries.

Innovation researchers have identified several typologies for classifying innovations in order to better understand them. One historical classification is product versus process innovations. *Product innovations* can be viewed as something new to an organization that is embodied in a physical object. An organization can be viewed as innovating if it invents the object, commercializes the object, or uses an object commercialized by others for the first time. *Process innovations* can be defined as value-adding changes to the set of activities necessary for the creation or distribution of a product or service. The use of automated production equipment or computers are easily understood examples of process innovations that change the way the product is made or service provided without necessarily changing the underlying product or service.

Another fairly intuitive but still subjective classification is radical versus incremental innovations (Marquis 1988). An innovation is considered to be radical if the improvements in cost or performance over existing products or services are dramatic, usually through the use of materials or systems that are very different from existing materials or systems. *Radical innovations* are said to be game changers because previous industry leaders can quickly be dethroned. For example, flash drives (also call thumb drives) made Iomega's zip drives obsolete because flash drives do not require a separate bulky device to read the media. *Incremental innovations* are considered to offer only modest improvements in cost or performance, so their diffusion may be slow. Because radical innovations can often support a significant price premium and can quickly achieve large market share, one might conclude that good leaders should always encourage their employees to pursue radical innovation. However, radical innovations require substantially more investment and are typically much riskier to pursue.

Slaughter (1998) applied two additional types of innovations identified by Henderson and Clark (1990) to the EPC industry. *Modular innovations* (not related to the modular buildings the reader is probably thinking of) are innovations that change one component of a finished product or one step of the entire process, but do not significantly affect other components or steps. Conversely, *architectural innovations* (again, not the architecture that may come to mind), significantly affect other components or steps, which likely require adjustments and negotiations with the affected parties.

Importance of innovation

In a classical economic treatise published in 1942, economist Joseph Schumpeter discussed how innovation within individual firms provides the *creative destruction* that directly affects the success of the firms relative to their competitors and leads to economic growth within society. Schumpeter and others have shown that firms that successfully manage innovation achieve increased productivity and/or superior products and services and push non-innovators out of business. Since then, economic researchers have analyzed the inputs (i.e. innovation-related expenditures) and results of innovation activities both for overall industries and for individual firms. Innovation-related expenditures include expenditures on "all scientific, technological, organizational, financial and commercial steps which actually lead, or are intended to lead, to the implementation of innovations" (OECD/Eurostat 2005: 18). Four sets of innovation benefits, i.e. drivers of innovation, are commonly cited:

- Decreased costs
- Increased quality and performance
- Increased customer and employee satisfaction
- Spillover effects throughout the organization

It may be helpful to examine each of these benefits, using examples from other industries as well as the EPC industry.

Decreased costs

As a result of innovation, many products can now be manufactured and services delivered today at a much lower cost in terms of real dollars. In other words, innovation has led to productivity gains (Schmookler 1966) and to products that provide comparable performance but cost much less, when adjusted for inflation, than products cost decades ago. Examples from the consumer world include computers, cameras, and music and video players. Examples from EPC include just about every building material used today. Some cost reductions have been achieved through new materials, such as using polymers instead of metals or wood. Other cost reductions have come through the use of innovative tools (from nail guns to laser leveling devices) and equipment (including laser- and GPS-enabled earthmoving equipment). While product innovations are most visible to many people, cost reductions through process innovations have been nearly as dramatic. Computers and networks have reduced the cost of communication and the transfer of information in general to pennies a day. (Consider how much construction companies used to spend on postage, duplicating, blueprint copying and so on.) Managers may not hear the word *reengineering* as much as they did during the 1990s, but applying information technology to restructured process steps still can lead to dramatic efficiency improvements, especially in EPC. Prefabrication and modularization are examples of both product and process innovations in EPC that can save money. Reduced costs, of course, can lead to higher profit margins or allow reduced prices that lead to increased sales and market share.

Increased quality and performance

Higher profit margins can also be achieved by delivering a product or service that customers are willing to pay more for than for competitors' products and services. *Differentiated* products and services are those that cannot be commoditized. Apple, for example, can charge more for an iPod than other firms can charge for generic MP3 players because the iPod is perceived as a "cooler" and better designed product. Consumers who buy Heinz ketchup will pay more for it than for generic ketchup because they believe Heinz ketchup tastes better, has better texture, etc. McKinsey or the Boston Consulting Group can charge more for their consulting services than can less reputable consulting firms offering the same services.

Innovations in EPC also offer organizations opportunities to differentiate themselves by offering higher quality or performance. With regards to design, Frank Gehry's clients do not contract with him because they know his design will cost less to design or construct than will designs by other architects. Engineering firms who have moved from 2D CAD to 3D solid modeling can demonstrate to clients how the latter system will enable conflict checking that will save construction time. With regards to construction, engineered wood products (e.g. roof and floor trusses, wood I joists) provide better dimensional stability, strength and stiffness than the sawn lumber products they replace (Toole 2001). The higher energy efficiency that customers in all construction market segments are increasingly demanding can be achieved by adopting innovative products and assembling them per an innovative design.

As is true for decreased costs, processes re-engineered and retooled through information technology can be a means of differentiation. For example, clients often value real-time transfer of information, including progress reports and images of progress to date. The use of integrated project software can give clients more confidence that submittals, requests for clarification and change order requests are being processed expeditiously.

Increased customer focus and employee satisfaction

It should be clear from the examples provided that innovation that leads to better quality and/or performance should lead to increased customer satisfaction. Less obvious is that innovation often leads to increased employee satisfaction as well. Many employees want to be creative, to be part of an organization that is considered forward-thinking, and to use their insights to better solve problems, to improve organizational processes and to satisfy customers. Conversely, employee satisfaction is typically lower in organizations in which opportunities for fresh thinking are few, where managers do not appreciate employees' suggestions for improvements, or where mediocre customer satisfaction is acceptable.

Spillover effects

Less obvious and measurable are the indirect effects from seeking innovation. As will be discussed later in this chapter, successful innovation requires an intentional focus on organizational goals and customer needs, being open-minded to experimentation and considering new ways of doing routine activities, an outward-looking and forward-thinking perspective, and a willingness to collaborate with internal and external groups. These organizational characteristics often lead to better formulation and execution of strategic goals, superior cost and quality management, and improved customer service even if innovation is not involved.

Innovation in EPC

Innovation researchers often view the EPC industry as being extremely non-innovative relative to other industries for three reasons:

1. EPC has historically had one of the lowest expenditures on internal R&D;
2. the rate of diffusion of innovations is typically slower than in many industries; and
3. EPC organizations rarely track innovation expenditures or results.

The CII innovation research indicated that EPC professionals tend to agree. Only 26 percent of the 150+ respondents indicated the EPC industry is mostly or highly innovative. Significantly more respondents (52 percent) indicated the industry is mostly or highly *non*-innovative. This same research confirmed that EPC organizations do not track innovation expenditures or results.

While the rate of innovation within EPC may indeed be slower than in other industries, EPC professionals know from their own experiences that innovation does occur within the industry. Table 5.1 opposite summarizes the results of the respondents' answers to "What do you think drives innovation in your capital projects organization? (check all that apply)."

While innovation does occur in EPC, research literature and our own observations indicate that innovation is almost always incremental, not radical. Some outsiders have concluded that EPC professionals must be irrationally clinging to current methods and materials and unwilling to learn or try new things. This may sound plausible given that the EPC industry (more than most industries) relies on professionals with highly specialized knowledge, be it in design or construction. An innovation, through its creative destruction nature, could make an individual person's or firm's expertise less valuable and lead to loss of wages, informal power and prestige, profits and market share.

Table 5.1 Factors that drive innovation in EPC

Reason	% checked
Reduce engineering and/or construction cost	72%
Increase our profit margins	63%
Serve our clients better	57%
Reduce engineering and/or construction duration	52%
Individual ingenuity and/or passion	49%
Achieve higher performance of the constructed facility	43%
Improve construction safety	29%
Limited resources require innovation	24%
Owners demand innovation	23%
Be more environmentally sustainable	20%
Global competition	16%
Subcontractors propose innovations	14%
Vendors propose innovation	12%
Improve the supply chain	9%
Other	3%

While fear of diminished economic and informal power cannot be ruled out as influencing innovation in EPC, many researchers perceive that the slow rate innovation in EPC can be explained *in part* by examining the *high number and complexity of barriers to innovation* that all EPC organizations face. These barriers include:

Unique projects

Most manufactured products and many service processes can be applied across a wide domestic or international base without modification. But every construction project is somehow unique in the characteristics of the site on which it is located, the footprint and appearance of the exterior, the interior space layout, and the interior finishes and HVAC systems chosen. Innovations that perform well on one project or in one organization may not perform well in different contexts (Toole 1998).

High project interdependence among diverse firms

EPC work product (buildings, plants, etc.) consists of many interacting parts and/or dynamic subsystems assembled by many different entities. Many innovations, especially radical ones, provide cost or performance advantages for one subsystem (such as wood framing) but affect one or more other subsystems or entities. For example, using composite steel/concrete decks may offer time and cost savings, they also affect the installation of mechanicals (plumbing, sprinkler, electrical, telecommunications and HVAC systems). Some product innovations not only affect the economics of installation but also affect the performance of other systems in unpredictable ways. As such innovators must often work hard to ensure an innovation will be accepted by all designers, subcontractors, local building officials and clients.

Market fragmentation

Many industries have several firms with large market power. For example, when Microsoft, Apple or General Electric offer a new product or service, they typically "move the market" because consumer demand will likely be high. Essentially no EPC firms, however, possess such market power. Few firms have the market share or customer loyalty that will cause their innovations to be embraced simply because of the reputation of the innovator.

Resistance to initial and fixed costs

Most innovations require the innovator to spend money to research and/or implement the innovation. EPC innovations with high initial costs are avoided because the profit margins on most construction projects are so low that a few construction firms have the cash needed to invest in them. If the market is down and firms' backlogs are low, managers want to sit on cash to weather the downturn. If the market is hot and firms' backlog is high, all of a firm's working capital is needed to finance work in progress. Some innovations, especially ones that involve complex equipment or licensing, result in the firm increasing their monthly recurring costs. Increased fixed costs associated with some innovations are avoided because high fixed costs make it harder for firms to weather the extreme swings in demand that the industry faces every decade or so.

Although these barriers are real, non-trivial, and industry-wide, EPC leaders cannot point to them as reasons why they do not encourage aggressive innovation in their organizations. The statement preceding the list stated these barriers only explain *in part* why innovation is low in EPC. The CII research provided additional answers. Survey participants were asked to identify the key perceived barriers to innovation within their organization. The results are shown in Table 5.2 opposite.

   While these data seem to confirm the importance of some of the industry-wide innovation barriers discussed above, it is important to note many of the factors are organizational in nature, that is, at least partially reflective of actions taken by management. *In short, the survey data and interviews indicate the low level of innovation in many EPC firms is a result of managers who have failed to implement cultures, policies and programs to enable effective innovation within their organization.* Given the prominent role that innovation has played in nearly all other industries and could play in construction, it is critical that EPC leaders learn why and how to enable innovation in their organizations.

Organizational enablers of innovation in EPC

Given the focus of this book, it is appropriate to first ask what the leaders' roles are in establishing and maintaining these characteristics, and what attributes and perspectives must leaders have to enable innovation within their organizations? The short answer is that

Table 5.2 Key perceived barriers to innovation on capital projects

Reason	% checked
Schedules and budgets are too tight to take a chance on something new	70%
Lack of resources (including staff time)	61%
Owner clients do not recognize the value	53%
Lack of a firm strategy for innovating	41%
Requiring project innovation costs to be born solely by the project	39%
Lack of organizational structure to nurture and follow through	35%
Potential reward is outweighed by the risk	33%
Overly restrictive project specifications	29%
Lack of communication between project participants	29%
Too many players in the process	29%
Lack of trust between project participants	22%
Rigid top-down command and control hierarchy	12%

leaders must drive every aspect of innovation and must have the proper attitude and practical skills to do this.

With regards to the proper attitude and skills, it is critical that leaders understand and manage two "big picture" aspects of innovation management. First, leaders must balance short-term goals with long-term goals. A manager who is maximizing the achievement of short-term goals and ignoring long-term goals will cut any expenditure that does not provide an immediate tangible payoff. For example, a short-sighted manager might freeze salaries for several years in a row in order to maximize short-term profits. As we all know, this would lead to reduced morale and productivity, talented personnel would start to leave the company and profits would decrease over the long term. Because innovations typically result from a year or more of research, experimentation and implementation (all of which consume cash, space and personnel resources), no innovation projects will be initiated if a manager is maximizing short-term profits. Leaders must acknowledge the achievement of long-term goals requires the investment of resources beginning immediately and continuing until the innovation has been permanently implemented within the organization. In short, leaders must take a long-term, holistic view of the innovation process (Oden 1997). Also, strategic clarity and consistency by leaders are important for sustained innovation (Rosenbloom and Cusumano 1988; Delphi Group 2006).

The second big picture idea is that innovation, probably more than any other process that organizations attempt, is highly uncertain. Managers who know how to perform sophisticated financial analysis to confirm that cash flows will provide an acceptable rate of return on capital equipment will likely be very uncomfortable with analyzing the expected payback of an innovation because the magnitude and timing of cash flows associated with both investments and revenues will be difficult to estimate. The quote from a respected European publication on innovation illustrates this well:

The decision to innovate often takes place under great uncertainty (Rosenberg, 1994). Future developments in knowledge and technology, markets, product demand and potential uses for technologies can be highly unpredictable, although the level of uncertainty will vary by sector, the life cycle of a product and many other factors. The adoption of new products or processes or the implementation of new marketing or organizational methods are also fraught with uncertainty. Furthermore, the search for and collection of relevant information can be very time-consuming and costly.

(OECD/Eurostat 2005: 30)

In addition to the leadership attributes just discussed, the CII innovation research team identified eight additional organizational attributes as being necessary for successful and repeated innovation. These attributes are discussed briefly in the following pages.

Innovation is a foundation of the organization culture

Innovation must be a core value within the organization and must underlie key organizational goals and processes. Norms (expected behaviors that are implicit) must be in place that encourage and facilitate innovation. These norms should include:

1. a focus on continual creativity, questioning the status quo (Sawhney and Wolcott 2004) and an openness to new ideas in all areas of the organization;
2. a recognition that a Not Invented Here syndrome cannot be tolerated (Chesbrough 2003, *Economist* 2007);
3. an emphasis on individual initiative and recognition that the entire organization must be committed to and involved in innovation activities (Daniel 2007);
4. a recognition that innovation can encompass or be manifested in all aspects of the organization's activities (Chesbrough 2003);
5. a greater emphasis on risk taking and a stated tolerance for greater risk; and
6. a focus on engagement throughout the organization and supply chain.

The CII innovation research team found that many of the individuals interviewed or surveyed perceive that innovation is valued within their organizations, but important foundational elements for innovation such as openness to new ways to do things and to taking short-term risks for potential long-term gains were missing. The EPC industry includes large and successful organizations that have large databases of past projects and a large repository of standard operating procedures, often associated in some way with Best Practices. Both past project databases and commitment to standardized procedures are double-edged swords. They can help large organizations consistently perform well on large and complex projects, but they also serve as inertial mechanisms to organizational change and barriers to the introduction of innovative products and processes. Innovative firms look forward as well as backward. They consider new processes as well as tried and true processes.

Budget allocations must be made in support of innovation

Innovation requires organization support. An integral and foundational component of this support is an explicit allocation of budget resource to innovation pursuits. The following budgeting principles should be considered.

1. Predetermined project budgets cannot accommodate innovation-related costs that were not considered when the budgets were established. Even in firms that have project funds set aside for innovation-related activities, it is appropriate to have corporate funds available to pursue corporate-wide innovation activities.
2. Corporate budgets should be allocated to identify innovations from outside the firm (including outside EPC) that could be implemented within the firm.
3. Once an innovation has been proven successful on one project, corporate budgets should be made available to diffuse the innovation to other projects.

Staff allocations must be made in support of innovation

Just as it is necessary to explicitly assign funding to innovation activities, it is necessary to explicitly assign staff to pursue innovation activities. The following staffing principles should be considered:

1. individuals should be tasked with identifying innovations from outside the organization that might be applied within the organization;
2. individuals should be tasked with facilitating implementation innovations on individual projects; and
3. individuals should be tasked with collaborating with individuals outside the organization (e.g. clients, subcontractors and vendors) to pursue joint innovations.

The clear picture that emerged from the dozens of interviews and hundreds of surveys conducted by the CII innovation research team is that EPC is a "lean, mean constructing machine," in which no one has the time to spend on innovation-related activities because they are so busy fulfilling their primary operational duties. When the team's research began in late 2006, every employee was busy trying to help their CII employer work through the organization's huge backlog. By the conclusion of team's research in mid-2009, the industry had downsized significantly due to the global economic downturn and most employees were trying to perform the duties previously performed by a subordinate or co-worker. What did not change over this period was that EPC managers were neither explicitly assigning innovation responsibilities to their employees nor ensuring employees had sufficient time in their day to pursue innovation-related activities.

Processes need to be put in place to support innovation

There has been a significant focus over the past two decades on managing organizational processes. Examples include total quality management, Six Sigma, earning ISO 9000 and 14000 designations, continuous improvement, and lean production. The reader will not be surprised then by the finding that organizations must have repeatable innovation-related processes that employees understand and are able to follow. This element was found by the CII research team to be a statistically significant difference between firms that were self-reported to be innovative versus those reported to be less innovative. It should be noted that, due to the nature of innovation, these processes cannot be described or managed as tightly as other operational processes. Specific process principles that should be considered are listed below.

1. Repeatable processes need to be established relating to the identification, evaluation and implementation of innovation on project and corporate levels.

2.   Processes should facilitate creative thinking and decision making.
3.   Processes should identify and meet customer needs.
4.   Promotion and bonus pay should reflect employee involvement in innovation activities (Gambatese 2007), even if innovations are unsuccessful.

Without established innovation-related processes, employees will be too busy achieving traditional project and departmental goals to pursue innovation, or they will focus only on following standard operating procedures that have worked on past projects but may not be the most efficient and effective way to complete their current projects. Young employees need to believe they can be promoted in part due to their willingness to pursue innovation.

Collaboration needs to be enabled

It was discussed earlier that the process on an EPC project always involves many diverse but highly interdependent entities. It was also discussed that some innovations in EPC are classified as "architectural" innovations in that they affect more than one component or phase of a construction project. Many promising information technology-based EPC innovations, in particular, are highly architectural and only deliver their full benefits when adopted on a project by all project participants. Effective creation or adoption of both product and process innovations typically require collaboration between multiple project entities.

Processes must be established by the project owner to enable project participants to identify innovations well suited for a project before the project begins in order for project budgets and infrastructure to include the innovation. Ideas must be brought into the firm from outside the firm, including from researchers and consultants (Bossink 2004). Communication between organizations must occur to ensure the innovation is implemented efficiently and effectively. Ideally, many of the project participants should have long-term relationships enabled by frequent communication. All project participants must feel the other participants are committed to the success of the project, not just to the success of their own organization. This way, the necessary linkages within organizations and between organizations will enable the collaboration and trust that are critical for innovation (Oden 1997; Hamel 2006; Kanter 2006; Boston Consulting Group 2007).

The long-term linkages and trust described in the previous paragraph do not often occur in EPC for various reasons. Leaders must establish appropriate organizational structures and infrastructures (typically information technology-based to supplement face-to-face meetings) to maintain these relationships. Perhaps the most important project factor enabling the necessary collaborative processes is primarily controlled only by the owner: contract type. The traditional design-bid-build delivery order method typically does not enable the collaboration that is needed for innovation. Design-build and integrated project delivery, especially when they involve co-location of design and construction personnel during the design phase, are much more conducive to collaboration (Gambatese 2007).

A new risk perspective needs to be adopted

One of the strongest recommendations of the CII innovation research team was the need for EPC leaders to change their risk perspective with regards to innovation. Look back at Table 5.2, which reported the key perceived barriers to innovation on capital projects from the CII innovation research team's survey. The fact that 70 percent of respondents indicated that

"Schedules and budgets are too tight to take a chance on something new" and 33 percent of respondents indicated "Potential reward is outweighed by the risk" were among their top seven barriers indicated that most firms lack the proper risk perspective on innovation. Innovation is viewed as simply too risky rather than as a necessary activity with inherent risk that must be effectively managed. As mentioned earlier, understanding that innovation involves a high level of risk and ambiguity is essential for successful innovation. Accepting and managing this risk requires a change in perspective from risk aversion to risk management. Two innovation risk principles should be considered:

(1) Shift from "single event" thinking

Leaders should not decide whether to pursue an innovation on a project based on the potential benefits and risks just to that one project. Instead, innovation decisions should be made using a multiple-project perspective. The first project on which an innovation is used will face high implementation and learning costs and high risk of failure yet only offer a modest set of benefits. On second and third projects when the innovation will be used, the costs and risk of failure will be much lower and the full benefits can be achieved. When costs, risks and benefits are amortized over multiple projects, pursuing an innovation will be viewed as a rational decision. Perhaps the casino analogy will be helpful to illustrate this point. Each individual table (e.g. blackjack) in the casino faces a risk that it could lose the casino's money over the course of an evening or even a week. But casino owners know that the risk of losing money over all tables in the casino over a night are very low because the odds are set up in their favor.

(2) Indemnify project managers who try an innovation

Because a project manager who first tries an innovation on his or her project will accrue nearly all of the costs and risks yet few of the potential long-term benefits to the organization, he or she will almost always decide not to innovate on his or her project. Yet, as just discussed, this decision will not prove optimal to the firm over multiple projects. At the least, such innovative project managers should not be penalized if their budgets or schedules are negatively impacted by trying a promising innovation. In addition, leaders should consider establishing salary, bonus and promotion mechanisms that will encourage individual project managers to incur risk by trying an innovation on their project that may reap rewards over multiple projects. Using the casino example again, individual dealers would likely refuse to deal a table if their own personal money was at stake because they could lose their life savings in one bad night. But casinos understand that individual tables can have bad nights so they let the dealer use the casino's money and do not fire a dealer whose table does not bring in expected revenues each and every night.

Manage organizational learning

Various innovation researchers have written that organizational learning and knowledge management are critical for sustained innovation (Rosenbloom and Cusumano 1988; Oden 1997; Delphi Group 2006; Boston Consulting Group 2007). The single event versus multiple event principle just discussed helps illustrate why this is true. If the project manager who first tries an innovation does not pass on the insights he gained about the innovation onto the project managers who will use the innovation on their projects, the high costs and risks and low benefits experienced by the first project manager will also be experienced by subsequent project managers. A key to innovation is experimentation and subsequent learning. Organizational learning must be established as an organizational value (Chinowsky and Carrillo 2007). Norms and processes must be established that encourage and enable candid sharing of lessons learned between project managers, especially when innovation is involved (Gambatese 2007).

Customer focus

It was discussed at the beginning of this chapter that some innovations reduce costs while others increase quality and performance. While the former can lead to improved profit margins, the latter are typically more important because they better satisfy customer needs and therefore can better improve profit margins, market share and customer loyalty. Organizations that appropriately focus on customers for innovating are known to:

1. emphasize learning the customer's business in an attempt to identify innovations that will help the customer;
2. effectively integrate customers in the development of the project or the product;
3. anticipate the future needs of customers by looking beyond current requirements (e.g. technologies, processes, etc.); and
4. focus on satisfying the end user more than in reducing costs.

Understand current status

The previous elements outline the requirements for establishing an innovative culture in the organization. However, prior to embarking on an aggressive change policy to put in place this culture, it is critical to understand where the organization is at the current time. To assist in this process, the authors created the Innovation Maturity Model (IMM) to assist organizations in determining current levels of innovation and providing recommendations for enhancing innovation in specific areas. The Innovation Maturity Model provides an organization with an evaluation of current innovation practices, an analysis of the areas with the greatest improvement potential, and recommendations for achieving this potential. The IMM is now an Excel-based tool that is available for any organization to evaluate, promote, and enhance innovation within their organization (Toole 2010).

Strategic innovation leadership

As detailed in the previous section, a successful innovation program faces diverse organization barriers. However, these barriers can be significantly reduced if the organization leaders provide a strategic commitment to an innovative culture. Put simply, top management must be sincerely and visibly committed (i.e. serve as champions) to innovation (Oden 1997; Bossink 2004; Boston Consulting Group 2007). What is the key to providing this strategic leadership? What is the key lesson that leaders need to draw from the discussion in this chapter? The answer is a commitment to organization learning and strategic vision that places innovation at the forefront of long-term organization success.

   The fundamental basis for establishing this strategic learning perspective for innovation is the requirement to enhance learning within the organization (Chinowsky and Carrillo 2008). Learning is categorized based on when and why it takes place and the effect that it has on those who are learning. The first type of learning can be thought of as incremental or reactive learning in which knowledge is gained in a piecewise manner as it becomes a necessity, while the second is a dynamic process of continual learning in which knowledge is proactively sought out before it becomes a necessity. In today's dynamic business world, it is no longer enough for organizations to emphasize the first type of learning. Rather, to enhance innovation, knowledge needs to be shared and utilized at an organizational level if a company hopes to survive. Hendriks and Vriens (1999) suggest that the knowledge assets possessed by a company create the possibility for sustainable competitive advantage. This being the case, a learning organization actively adapts individual knowledge into innovations that can be readily used to benefit the organization as a whole.

The second strategic imperative for establishing a successful innovation program is the need to implement new practices on a regular basis. The need to enhance an organization's productivity and effectiveness in response to changing environments is a recurring theme through business literature in particular and all areas of production in general. As outlined by Collins (2001), the greatest threat to achieving greatness is an organization that is content with being very good at what it does. In the context of innovation, this thought transfers to the greatest threat to achieving an innovation culture the being content with established practices that have achieved acceptable and above average results.

In summary, learning and collaboration are central themes in innovation. Collaboration is a central requirement for implementing new innovations. These innovations are a key to continuously updating an organization's approach to the market forces impacting business practices. Given this relationship, organization leaders need to place a greater strategic priority on knowledge sharing as a path to increasing the introduction of new innovations. Once again, it is a strategic perspective on organization performance that brings innovation to the forefront of long-term performance concerns, and it requires sincere organization leadership to bring this perspective to fruition.

Conclusions

EPC owners focus on bringing capital projects in on time, under budget and without injury. EPC contractors pride themselves on making their customers happy by focusing on these same goals. Both sets of organizations believe the best way of achieving these goals is by implementing best practices, that is, by adopting what has worked on past projects and eliminating all unnecessary sources of risk to achieving project goals. Both sets of organizations are lean, mean design and/or construction machines, with nary an ounce of organizational fat. The vast majority of EPC organizations are rightfully proud of their critical abilities to manage successful capital projects.

The trouble is that these project management strengths are exactly what make the organizations mostly inept at being innovative. EPC organizations like the idea of being innovative but most of them are poor at identifying, nurturing, testing, and managing knowledge regarding innovations within their organizations. New methods, processes, products, and systems may sound exciting, so the thinking goes, but they are too costly and too risky to pursue even on the smallest projects. And who has the time and motivation to investigate potential innovations anyway?

For decades, EPC organizations have given lip service to innovation but secretly pointed at a plethora of factors—all external to their organization—that have kept them from being innovative. But today's volatile and global markets, today's ever increasingly complex capital projects and tomorrow's workforce will no longer allow EPC organizations to ignore opportunities for innovation. EPC organizations who continue to focus solely on achieving short-term project goals at the expense of developing long-term capabilities to manage innovation and change will find themselves following the same path of the US shipbuilding industry: decreasing productivity, decreasing profit margins and decreasing market share.

The authors of this chapter believe every EPC organization has a significant opportunity to improve its innovation performance. However, increasing innovation will require a dramatic shift in risk perspective by organization leaders and establishing organizational-wide culture that embraces new approaches. It will require a commitment to repeatable innovation processes and resource allocation. It will require major changes in contracts and delivery order methods to enable the collaboration needed for innovation. None of these critical organizational attributes will be easy to establish, but we are confident the payoff will be worth the effort.

References

Bossink, B.A.G. (2004), "Managing Drivers of Innovation in Construction Networks," *Journal of Construction Engineering and Management*, 130(3): 337–345.

Boston Consulting Group (2007), "Innovation 2007: a BCG Senior Management Survey," online, available at: www.bcg.ch/fileadmin/media/pdf/innovation_2007.pdf.

Chesbrough, H. (2003), "The Era of Open Innovation," *MIT Sloan Management Review*, 44(3): 35–41.

Chinowsky, P.S. and Carrillo, P. (2007), "The Knowledge Management to Learning Organization Connection," *Journal of Management in Engineering*, ASCE, 23(3): 122–130.

—— and —— (2008), "A Strategic Argument for Knowledge Management," *Proceedings of the Third Specialty Conference on Leadership and Management in Construction*, Lake Tahoe, CA, October, 2008, CIB.

Collins, J.C. (2001), *Good to Great*, New York: Harper Business.

Daniel, D. (2007), "Five Steps to Managing Innovation," online, available at: www.cio.com/article/125700/Five_Steps_to_Managing_Innovation.

Delphi Group (2006), "Innovation: From Art to Science," online, available at: www. delphigroup.com/about/pressreleases/2006-PR/20060331-innovation.htm.

Dikmen, I., Birgonul, M.T., and Dikmen, S.U. (2005), "Integrated Framework to Investigate Value Innovations," *Journal of Management in Engineering*, 21(2): 81–90.

Economist, The (2007), "Lessons from Apple," June 9, 383(8532): 11.

Freeman, C. (1986), *The Economics of Industrial Innovation*, second edition, Cambridge, MA: MIT Press.

Gambatese, J.A. (2007), *Innovation Manual of Practice: Energizing Innovation in Integrated Project Delivery*, Research Report submitted to Charles Pankow Foundation and Design-Build Institute of America, online, available at: www.spur.org/files/pankow/01-06IMOP-Dec2007.pdf.

Hamel, G. (2006), "The Why, What and How of Management Innovation." *Harvard Business Review*, February, online, available at: www.netscope.com/pdf/Management Innovation.pdf.

Henderson, R.M. and Clark, K.B. (1990), "Architectural Innovation: The Reconfiguration of Existing Product Technologies and the Failure of Established Firms," *Administrative Science Quarterly*, 35(1): 9–30.

Hendriks, P.H.J. and Vriens, D.J. (1999), "Knowledge-Based Systems and Knowledge Management: Friends or Foes?" *Information and Management*, 35(2): 113–125.

Kanter, R.M. (2006), "Innovation: The Classic Traps." *Harvard Business Review*, November, online, available at: http://hbr.org/product/innovation-the-classic-traps/an/R0611C-PDF-ENG.

Marquis, D.G. (1988), "The Anatomy of Successful Innovations," in M.L. Tushman and W.L. Moore (eds.), *Readings in the Management of Innovation*, New York: Ballinger Publishing Co.

Oden, H.W. (1997), *Managing Corporate Culture, Innovation and Intrapreneurship*, Westport, CT: Quorum Books.

OECD/Eurostat (Organisation for Economic Co-operation and Development and Statistical Office of the European Communities) (2005), *Oslo Manual: Guidelines for Collecting and Interpreting Innovation Data*, online, available at: www.oecd.org/document/33/0,3343,en_2649_34273_35595607_1_1_1_37417,00&&en-USS_01DBC.html.

Park, M., Nepal, M.P., and Dulaimi, M.F. (2004), "Dynamic Modeling for Construction Innovation," *Journal of Management in Engineering*, 20(4): 170–177.

Rosenbloom, R.S. and Cusumano, M.A. (1988), "Technological Pioneering and Competitive Advantage: the Birth of the VCR Industry," in M.L. Tushman and W.L. Moore (eds.), *Readings in the Management of Innovation*, New York: Ballinger Publishing Co.

Sawhney, M. and Wolcott, R. (2004). "The Seven Myths of Innovation," *Financial Times*, September 24.

Schmookler, J. (1966), *Invention and Economic Growth*, Cambridge, MA: Harvard University Press.

Schumpeter, J.A. (1942), *Capitalism, Socialism and Democracy*, New York: Harper.

Slaughter, S. (1998), "Models of Construction Innovation," *Journal of Construction Engineering and Management*, 124(3): 226–231.

Toole, T.M. (1998), "Uncertainty and Homebuilders' Adoption of Technological Innovations," *Journal of Construction Engineering and Management*,124(4): 323–332.

—— (2001), "The Technological Trajectories of Construction Innovation," *Journal of Architectural Engineering*, 7(4): 107–114.

—— (2010), *Innovation Maturity Model Tool*, CII RT 243 Expanding and Enhancing Innovation Research Team, online, available at: www.facstaff.bucknell.edu/ttoole/Innovation/.

6 Human resource development: rhetoric, reality and opportunities

Andrew Dainty and Paul Chan

Introduction

Of all of the functions vital to ensure the responsiveness and agility of the contemporary construction firm, it is perhaps the human resource development (HRD) function which is the most crucial. The emergence of HRD as a strategic activity arguably reflects changes in the human resource management (HRM) new orthodoxy. The HRM literature has moved from being a largely reactive philosophy (impacted upon *by* policies) to become a more proactive function (see Bellini and Canonico 2008). This is a clear acknowledgement that the way in which people are managed affects employees' commitment to improving the firm's products and services (see Marchington and Wilkinson 2005). Seen in this way, HRM can have a strategic influence within many organizations and can be viewed as a route to sustained competitive advantage. Such an assertion is supported by analysis of firms' financial performance which suggests that high performance work practices are positively linked to firm performance (Huselid 1995), although this is by no means uncontested (Purcell 1999).

In considering the role and importance of HRM to the performance of the modern construction firm, the role of HRD as a key influence on competitiveness becomes especially important. This is because the role of learning and development is so central to firms wishing to maintain their competitive position. As will be explored within this chapter, all firms must evolve their capabilities in order to exploit market opportunities and to remain responsive to the changing conditions within which they operate. Thus, and notwithstanding difficulties of defining HRD that are discussed later in this chapter, HRD can be seen to represent the 'developmental' aspects of the strategic human resource management cycle in seeking to align the development of people with the desired direction of the business (Loosemore *et al.* 2003). This is not to infer that normative models for how to manage human resources exist which should be universally adopted, but that HRM should be intertwined with the strategic management of the firm.

In this chapter we explore the role and importance of HRD within the context of the contemporary construction organization. We examine this in relation to four key questions. First, we explore *what HRD is*, examining the genesis of the concept and its theoretical underpinnings within the mainstream business and management literature. In doing so we reveal its inherent ambiguity given differing traditions and ontological perspectives which have shaped it. We bring this into practical focus by examining the manifestation and position of HRD within current HRM practice within the construction sector. Second, we explore *the importance of HRD in construction and the disconnections between rhetoric and reality* of HRD in practice. The particular focus here is on the structure of the industry and how this shapes the enactment of HRD in construction. We emphasise the importance of taking into account the way in which construction activity is organized and delivered. Having established the role and importance of HRD and the barriers to its enactment within the sector, we then examine ways in which it can be managed within the contemporary construction business. In this section we examine *how we can do HRD in construction* in order to support and augment a broader organizational learning climate, and connect this with more strategic concerns of the construction firm. Here we espouse the virtues of developing into a reflective 'learning organization', but also the need to consider stakeholders external to the firm when considering HRD approaches. This need for a more 'externalised' view of the function is crucial if it is to be seen as more strategically important in construction. Finally, we briefly examine potential *future directions for HRD*, suggesting further areas of development in relation to research, business policy and practice which need to be taken in order to begin to address the challenges set out elsewhere within this book. Again, the intention here is not to provide a normative framework for research or practice, but rather to suggest some of the issues likely to challenge construction organizations in the future.

Our treatment of HRD and associated concepts has necessarily required us to consider its enactment in context. This inevitably requires us to take account of existing industry structures and the ways in which construction organizations tend to operate. In considering HRD in this way, two important principles undergird the perspective adopted within this chapter. First, we deliberately adopt a *critical perspective* in attempting to understand the HRD function, and its role and status within the contemporary construction organization. Whilst we acknowledge the performative emphasis of HRD (in recognition of the wider objectives of this book), we also argue for a higher priority for the function as a key enabler of employee well-being and a positive employment relationship, as well as for supporting the performance of the firm. In this way we attempt to eschew some of the tensions which have emerged in reconciling perspectives on learning and development with the performance of the firm. Such a perspective is consistent with calls for a stronger focus on corporate social responsibility (CSR) (Murray and Dainty, 2009). A second underlying principle of our treatment of HRD within this chapter is that we have deliberately attempted to avoid the temptation to offer prescriptive advice in terms of how to better develop policies or implement HRD as a performance. As will be seen later, such models are inadequate for recognising the individual requirements of both construction organizations and the individuals affected by HRD strategies.

What is human resource development? Theoretical underpinnings and definitions

Constructing a precise definition of HRD is extremely problematic, and may even be counterproductive in that it would narrow down a broad set of interrelated activities which necessarily change in response to dynamic environmental influences. Nevertheless, it is important to chart the trajectory of the concept, exploring its interrelationship with the broader evolution of HRM and related strategy literatures before exploring its application in practice. In order to understand the theoretical provenance of HRD, it is important to position it in relation to the concurrent development of the literature on competitiveness, a literature which has been highly influential for those researching firm performance within the construction sector. Such an understanding will reveal how a more strategic view of the role of HRM emerged from the 'Resource Based View' (RBV) of the firm (c.f. Barney 1991), and how later theories around dynamic capabilities (c.f. Teece *et al.* 1997) have posed challenges for the HR function to ensure organizational agility in response to constantly changing requirements.

The Resource Based View of the firm (RBV) emphasises endogenous factors as the source of competitive advantage (Hoskisson *et al.* 1999 cited in Wright *et al.* 2005). In doing this, the RBV has been the most influential theory of competitiveness in terms of shaping development of strategic human resource management (SHRM), especially since SHRM tends to draw heavily on the concepts and theories from this broader strategy literature (Wright *et al.* 2005: 17–18). Wright *et al.* point out that of particular significance is that the shift to the RBV has raised the profile of people as the source of competitive advantage. This, in turn, raises the importance of concepts which support the development of people which now form the central components of HRD as will be explored below.

Although the original articulation of the RBV is generally attributed to Wernerfelt (1984), Barney's (1991) description of the firm-level characteristics which enable competitive advantage has provided the touchstone for developments in this area. Essentially, Barney's thesis was that resources can provide competitive advantage if they are rare, valuable, inimitable and non-substitutable. Seen in this way, it is the firm's development and positioning of these resources which lead to a source of sustainable competitive advantage. The RBV's focus on internal resources has ensured its enduring popularity in terms of shaping the SHRM literature (Wright *et al.* 2005). Moreover, it has provided a powerful lens for understanding construction organizations and the ways in which they position resources to strengthen their competitive position (see for example De Haan *et al.* 2002).

A potential weakness of the RBV is that it assumes that capabilities will endure over time. Clearly, for project-based organizations like construction firms that tend to focus on developing client-oriented products and systems, they will continually build capability through the involvement with the product. In other words, the learning that accrues when a firm moves into a new market can be understood as a dynamic process of building capability over time (see Brady and Davies 2004). Thus, as Davis and Walker (2009) suggest, construction organizations require agility in terms of delivery and knowledge of what is desirable. This requirement is reflected by Teece *et al.* (1997), who suggested that organizations need to develop their capabilities in order to respond to dynamic environments, a requirement which has been coined '*dynamic capabilities*'. They further suggested that:

> The competitive advantage of firms is seen as resting on distinctive processes (ways of coordinating and combining), shaped by the firm's (specific) asset positions (such as the firm's portfolio of difficult-to-trade knowledge assets and complementary assets), and the evolution path(s) it has adopted or inherited.
>
> (Teece *et al.* 1997: 509)

In other words, dynamic capabilities comprise organizational routines through which firms are able to reconfigure their resources in response to market change (see Eisenhardt and Martin 2000).

Despite its rhetorical attractiveness, the concept of dynamic capabilities has not had as significant an impact within the construction management literature as the RBV. However, recently Green *et al.* (2008) explored its relevance for understanding the strategy of construction firms, concluding that whilst it reflected the readjustment made in response to changing environments, it remains an empirically elusive and tautological construct. This notwithstanding, it has important implications for HRD in that learning and skill acquisition become key contributions to sustaining competitiveness (Teece *et al.* 1997). This 'knowledge-based' view of the firm (Grant 1996), which further emphasises the role of the individual as the primary actor in the creation of knowledge, adds additional weight to the role and importance of HRM in shaping the way in which knowledge is managed within the firm. Considered in the project context of construction, the dynamism of such capabilities and knowledges can be seen to be particularly apparent. Indeed, Davies and Brady (2000) suggest that 'project capabilities' emerge within firms involved with delivering complex products and systems (CoPS). Such capabilities are manifested in the preparation, development and life-cycle implementation and management of customer-specific solutions. Thus, learning which accrues through and from projects can be seen as key to competitiveness within such organizations (Brady and Davies 2004).

The discussion to date has emphasised the crucial importance of *learning* as a source of competitive advantage in order to maintain responsiveness and agility. Whereas in the past the development function has typically been associated with training, contemporary perspectives on the development function have shifted the emphasis to more of a learning perspective. Such a perspective is embodied in the concept of the 'Learning Organization' (c.f. Senge 1990), a powerful and alluring concept in the context of ensuring the ongoing strategic positioning of the firm. A 'Learning Organization' seeks to facilitate the learning of all its members, enabling it to transform itself in accordance with its operational requirements (see Pedler *et al.* 1991). This places a renewed emphasis on the development of people within organizations as the source of competitive advantage delivered through the HRD function.

In comparison with human resource management (HRM), human resource development (HRD) is a relatively new concept, and as such continues to evolve within practice and research (Sambrook 2004). It also remains a rather nebulous concept and, to some extent, one which evades clear definition. Sambrook (ibid.) suggests that one reason for this is that HRD can be explored from a variety of perspectives (e.g. from economics, psychology or systems viewpoints). This inevitably leads to a multiplicity of definitions and views on the construct, and to reinterpretations within different contexts. Thus, rather than attempt to define HRD as a homogenous activity, it is best understood as a label for a range of interrelated concepts and activities through which people are developed within organizations.

In recent years, views on HRD have tended to be polarised by a focus on either 'performance' or 'learning' (Simmonds and Pedersen 2006). As Simmonds and Pedersen discuss, whereas learning and development has provided the dominant discourse within the UK (Garavan *et al.* 1999), a performance orientation has predominated in the USA (see Sambrook 2004). However, considering the competitive pressures and structural form of construction companies, it is likely that even in the UK, where most construction activity is outsourced and training and development funded via a levy and grant system, performative motives are likely to dominate the HRD agenda. This trend in the UK has arguably led to a system whereby skills development is seen as an externalised activity, and any training which does go on within the firm must evidence performative advantages. Thus, despite the rhetorical attractiveness of HRD as a public relations vehicle for corporate social responsibility (c.f. Sambrook 2004), and whilst there are obvious HRD benefits for employees in relation to their personal development and employability, organizations tend not to promote learning for philanthropic reasons, but rather for performative ones (Marchington and Wilkinson 2008).

Importance of HRD in construction and disconnections between rhetoric and reality

And so what activities constitute HRD within the contemporary construction organization? The various elements of HRD include learning, education, development and training (Armstrong 2003). However, whilst terms such as education, learning and training are often used interchangeably, there are important distinctions between them (Marchington and Wilkinson 2008). Traditionally, HRD has been attached to formal *training*, instructor-led and often delivered away from the workplace (Sloman, 2005; cited in Marchington and Wilkinson 2008). *Education*, on the other hand, suggests the development of intellectual capability via the learning process; *learning* referring to the changes in skill, attitude and knowledge, and *development* as an overarching term which infers both training and learning over time (ibid.). In practice, a combination of these learning strategies will characterise the HRD process; people will learn as individuals and collectively as members of project teams and formalised training will draw upon the informal work-based experiences of learners and will be put into practice within their workplace environment. Within an experiential-learning based industry such as construction, specific project requirements will shape training requirements and will also afford specific learning opportunities.

A review of the construction HRM literature reveals that it is arguably a devalued function which has struggled to justify its role and importance within many construction organizations (see Druker and White 1995). The emergence of a more strategic view of HRM (SHRM) has, however, provided a clearer (albeit contested) link between the HR function to the performance of the firm (Loosemore *et al.* 2003). As a result, the role of human resource *development* has risen in prominence as a key enabler of organizational performance and, as will be explored later, as a key influence on competitiveness.

In attempting to develop a working definition for HRD within the construction context, it is also important to recognise the range of structural factors which influence the way in which HRD is managed and enacted within construction organizations. These include the propensity of construction firms to retain only a core group of knowledge workers (Dainty *et al.* 2007), the inherent requirement of such individuals to offer functional flexibility in the face of continual change (Dainty *et al.* 2009; Loosemore *et al.* 2003), and the ad hoc and largely unstructured ways in which HRM strategy tends to be enacted within construction organizations (Raidén and Dainty 2006). Hence, taking into account the multiplicity of different ways in which development occurs, and the nature of workplace environment within which it takes place, HRD in construction can be seen to reflect Simmonds and Pedersen's (2006: 123) definition of the function as: *"a combination of structured and unstructured learning and performance-based activities which develop individual and organisational competency, capability and capacity to cope with and successfully manage change."*

Thus, HRD can be seen as a part of key HRM practices underpinning competitive advantage, and as being squarely aligned with developing the endogenous capabilities implicit within the RBV and dynamic capabilities models. This renders HRD a rhetorically attractive device for construction organizations as will be explored below. The challenge remains, however, as to how construction firms operationalise and embed HRD practices. This is considered especially problematic given the structure of the industry.

The construction industry is generally criticised for its relatively low uptake of HRD (see Forde and MacKenzie 2007; Dainty *et al.* 2007). So, if skills development is such a good thing, why is there a disconnection between the recognition of the importance of HRD and the reality of a relatively low level of training participation in the sector? In the next section, an attempt is made to explain possible reasons for this perceived gap, pointing mainly to the evolving structure of what is a fragmented industry and the difficulties associated with the definition of skills in the sector.

Evolving structure of the industry

Globally, the output produced by the construction industry is sizeable, at approximately 10 per cent of many countries' GDP. When professional services are included in accounting for the sector, estimates of the contribution that the industry makes can often increase to around 20 per cent. Policymakers and academics have devoted much energy to defining what constitutes the construction industry, usually with a certain level of contention. Increasingly, boundaries between industries become fuzzy when one considers that the delivery of any construction product could encompass *inter alia*: the extraction of raw materials; the freight and transport of products that get assembled into a building; the manufacturing of components; and the professional services that support the design, construction and maintenance process (see Pearce 2003). It would seem therefore that the construction industry is no longer just about construction, but also includes goods and services produced by the logistics industry, manufacturing industry and the professional services industry. In fact, commentators have long acknowledged that defining the construction industry is fraught with problems and presents only artificial boundaries of analysis (see Groák 1994).

The complexity of defining what the construction industry is and does is further complicated by shifting trends in organizational structure of businesses that operate in the sector, which have arguably evolved to reflect the socio-economic context within which firms operate. In the past, construction companies used to be large outfits that hold many of the traditional functional departments (e.g. finance, personnel, sales and marketing, planning etc.) and employ a majority of the workforce directly including manual labour. Increasingly, however, a majority of companies associated with construction work in many countries tend to be small- and medium-sized enterprises (SMEs). Large construction companies no longer operate at the 'coal-face' of construction work, i.e. they no longer employ manual construction workers to do the work. Instead, many large construction companies act as 'system integrators', and merely manage the design and construction process and there is also a growing trend for these companies to be involved in the maintenance and operations of the built environment. The recruitment of manual construction workers is devolved to the SMEs as much of the work gets subcontracted through multilayer supply chains. Such a phenomenon has been documented as the 'hollowing-out' process of firms (see Castells 2002), and shifts the analysis of organizations away from the study of firms to the study of the networked organization (Grugulis *et al.* 2003; Kinnie *et al.* 2005; Grimshaw and Rubery 2005). Indeed, traditional firm boundaries are being disordered (see Marchington *et al.* 2009, 2010) and issues regarding who holds the responsibility for, and how firms in construction actually engage in, HRD become ever more blurred. Crucially, from a strategic management perspective, such a transformation implies a shift in focus in terms of what HRD means to businesses that operate in the construction industry. Specifically, *where* a company is placed along the supply chain determines the way in which HRD is conceptualised and operationalised; theoretically, companies that are placed higher up on the supply chain would tend to focus on enhancing their management capabilities whereas those further down the supply chain would stress on the development of manual craft skills.

Of course, the perpetuation of subcontracting has been known to result in piecemeal approach to work, and the pursuit of the profit motive coupled with the restricted capacity of SMEs to engage in training activities tend to limit the extent in which HRD is practised in the industry. Moreover, the construction industry is often used as a barometer of economic performance, and therefore is exposed to the vulnerabilities of economic cycles of boom and bust. Given such uncertainties, firms are less likely to engage in skills training and development, which requires a longer-term view (see MacKenzie *et al.* 2000). Furthermore, the industry epitomises the pinnacle of flexible organization and its deeply entrenched reliance on self-employment (Winch 1998) and contingent labour (Forde and MacKenzie 2007) reduces the industry's propensity to train. It is inevitable that the industry consistently reports a low average training investment of around one-person-day per year (see e.g. BERR 2007). So, the evolving nature of the industry – typified by the 'hollowing-out' of construction companies, the proliferation of subcontracting and the reliance on self-employment and agency workers – makes it more difficult for the nature of skills to be defined. Consequently, there is growing confusion as to what skills ought to be developed for the sustenance of what is an increasingly complex industry.

Difficulties associated with the definition of skills and the problematisation of HRD

At a very basic level, skills can be distinguished as being either specific or generic. Specific skills, as the name suggests, contribute specifically to a particular firm; whereas generic skills can be applied within any context. So, the ability of a worker to install a certain boiler system that a company specialises in can be said to be a specific skill, whereas team working and interpersonal communication skills can be said to be generic or transferable skills. Conventionally, the differentiation of the specificities of skills facilitate in helping policymakers decide on who pays for HRD. So, the logic follows that specific skills should be paid for by the firm that benefits from these, whilst generic skills should be paid for by the individual in whom the skills reside, or more commonly by the state through investment in the education system.

The distinction between specific and generic skills originated from Becker's (1964) theory of human capital, which relies on the rationalistic idea that one's sole purpose of survival is to maximise economic utility (Pearce 2006). Put another way, HRD efforts serve only to maximise one's productive capacity, which in turn results in maximisation of one's wage. Consequently, it is perfectly reasonable that one should invest in enhancing skills because one desires to become more productive. The concept of skills, under this economic perspective of human capital, implies that skills reside in workers themselves so that any increase in skills levels will bring about a corresponding rise in worker productivity and wages – the concept of wage labour.

The notion of skills is certainly complex. Drawing on Cockburn's (1983) writings, Grugulis (2003a; Grugulis *et al.* 2004) proffered three perspectives of skill:

> there is the skill that resides in the man himself, accumulated over time, each new experience adding something to a total ability. There is the skill demanded by the job, which may or may not match the skill in the worker. And there is the political definition of skill: that which a group of workers or a trade union can successfully defend against the challenge of employers and of other groups of workers
>
> (Cockburn 1983: 113; c.f. Grugulis 2003a: 4)

Accordingly, the first relates to the conventional economic perspective of human capital, and the latter two would be governed by a more sociological lens (Grugulis *et al.* 2004). Of course, Clarke observed that skills can mean very different things to the worker and the employer, as she noted,

> whilst training creates skills, these skills have different values for the worker who owns, sells, employs and attempts to conserve them than for the builder (employer) who buys and consumes them .... Under a capitalist mode of production ... the determination of training provision is only possible through an analysis of changes in production and in the social relations regulating the labour process
>
> (Clarke 1992: 6)

thereby integrating the perspectives of skills at work and the socially-constructed nature of skills definition between the actors concerned.

So, what are the implications of the complexities surrounding the definition of skills? For one, the economic perspective promoted by human capital economists like Becker (1964) is limited in attaining the ideal state of more skills leading on to greater productive capacity for both the firm and the worker. Instead, there is the tendency for employers to pursue a strategy of more general skills and less specific skills when a market becomes more competitive (see Groen 2006). This coincides with the suspicion that employers are inclined to abdicate from the responsibility of investing in skills development (Dainty *et al.* 2005) for fear that other employers would poach the workers once they have completed the training. There is mounting evidence that the shifting employer preference for general skills lends further support to the deskilling (Braverman 1974) of firm-specific skills. Recent work on British statistics has also suggested that the trend towards general skills has not escaped the construction sector. Interrogating the UK Labour Force Survey statistics between 1989 and 2005, Beaney (2006) investigated the nature of residential and sectoral migration in relation to construction and concluded that skills-specificity in construction has declined over the period of study. Arguably, the trend towards greater skills generality in construction is set to grow with the expansion of the labour market through the influx of migrant workers from overseas.

Grugulis *et al.* (2004) provide a more cynical outlook: the growing desire of employers to focus on generic skills offers on the one hand a false sense of upskilling among the workers, and on the other virtually no benefit in terms of wage premium. Becker's (1964) belief that investment in human capital would reap benefits of greater productive capacity and wage growth would therefore appear somewhat questionable. Indeed, recent evidence in the UK construction industry showed little correlation between skills levels and wage rates (Clarke and Herrmann 2004). For Grugulis (2003a: 5),

> the key issue here is not that technology, market forces or flexibility do not support skills development: it is that they do not inevitably do so. There are choices to be made about the ways that work is designed, monitored and controlled, and these choices will affect skills in a range of ways

Arguably, the economic perspective of human capital emphasises human resources as an economic factor of production and potentially plays down the human benefits that can be accrued through development (see also Nyhan *et al.* 2004).

Up to now, much of the writings have criticised employers for their unwillingness to invest in training and development. As was alluded to above, however, more recently sympathetic commentators have recognised the tensions inherent in construction companies having to juggle the short-term need for profitability with the long-term employee interests of skills development. Raidén and Dainty (2006) used the phrase 'chaordic organization' to describe how construction companies deal constantly with both the chaotic business environment and the orderly, strategic planning of skills. They recommended for more research to examine the characteristics of construction organizations as chaordic organizations. Chan (2007), in a single case study of a medium-sized specialist contracting firm, observed that the particular employer actually had quite a sophisticated, if informal, way to ensure that employee needs for personal development were met alongside the business need for expansion and diversification. This involved *inter alia* a traditional, family-business orientation that encouraged an 'open door' approach between line managers and workers so as to ensure that both the employer and employee needs are mutually satisfied on an individual basis (see also Chan and Kaka (2007) on the usefulness of family-business approaches to organizing construction).

It is perhaps worthy to note that specialist contracting firms increasingly have direct responsibility over the production process in construction, and so they would have a higher propensity to employ skilled manual construction workers directly. In the case study of the specialist contracting firm, Chan (2007) found that the firm tended to maintain an in-house workforce of skilled manual workers whilst sourcing the semi- or unskilled general labourers from elsewhere, e.g. employment agencies. Accordingly, specialist contracting firms would desire to continuously develop and improve the specific (technical) skills base of their workforce. Chan (2007) observed that specialist contracting firms would either invest in appropriate education and training courses, or poach talented individuals from rival firms that they interact with as they go about their day-to-day operations at the project site. Furthermore, it is not uncommon for firms to try workers out for a probationary period before offering stable employment to skilled workers. Such practices only illustrate the flexible labour market in which the construction industry tends to operate, and the deeply entrenched informality that form the modus operandi for HRM in construction. Nonetheless, the study of how such firms in construction actually go about HRD remains somewhat under-researched (see Raidén and Dainty 2006; Chan, 2007).

Thus, it can be seen that the complex and dynamic nature of skills, coupled with the diverse forms of organizations that operate in the construction industry, serves only to add confusion as to how firms can go about engaging in HRD in a strategic and systematic fashion. Indeed, given this structural context the capability of the firm to enact HRD can be seen as highly problematic. Commentators have suggested that HRD is best served by ensuring the strategic identification of what skills really matter, for both employers and employees, and enabling skills development and learning to take place through appropriate delivery mechanisms. In the next section, we put forward a number of ways in which HRD can be promoted in construction organizations, and critical considerations associated with enhancing organizational learning in construction.

How can we do HRD in construction?

Given the constraints that can inhibit the extent in which the industry engages in HRD discussed so far, it is now important to focus on what businesses can do to adopt HRD strategies and tactics effectively. As was explained in the first section of this chapter, HRD is a potentially good thing for individual workers, employers and society at large. However, as the preceding section has revealed, the ability of construction firms to embrace HRD practices is impeded by the industry structure within which they operate. The plurality of organizations and corresponding interests that exist in the industry, which results in divergent conceptualisations of skills, can prevent meaningful engagement of employers and individuals in HRD practices. In this section, the focus switches to what companies can do to mitigate the shortcomings described previously, with a view to improve HRD uptake in the industry. Two broad strategies are considered in turn. First, a plea is made for individuals, managers and workers alike, to constantly reflect on the organizational development needs and move towards becoming a learning organization that accrues benefits for all. Whilst reconfiguring the internal dynamics of HRD is critical (c.f. dynamic capabilities), it is increasingly important to acknowledge that exogenous influences also matter in HRD. Therefore, the section will also consider how companies can usefully engage with external stakeholders to boost its HRD efforts.

Internalising HRD: becoming a reflective learning organization

Organizational learning, as an academic concept, has gained much prominence since the 1990s. According to Burnes *et al.* (2003), scholars such as Argyris have been writing about organizational learning for decades. Yet, following Senge's (1990) publication of *The Fifth Discipline*, organizational learning has moved "from being a subject for serious academic study to a hot board room topic in the West", perhaps due to "the pace of change and the competitive threat posed by globalisation" (Burnes *et al.* 2003: 452). Indeed, organizational change, together with the global drive in moving up the value chain, the importance of the knowledge economy (particularly to the developed world), and the mantra of continuous improvement, have all mobilised research and practitioner interest in the role organizational learning can play in securing competitive advantage. Some would even go as far as to state that organizational learning presents the only form of competitive advantage (see e.g. Stata 1989; Kululanga *et al.* 2001).

However, the notion of organizational learning remains an abstract concept that practitioners can at times find difficult to operationalise. Huysman (2000), for instance, suggested that we have not fully understood the conceptual processes of organizational learning, whilst Lähteenmäki *et al.* (2001) observed that the reconciliation between individual learning and collective organizational learning has not materialised. Lipshitz *et al.* (2002) also reiterated that the concept, like many concepts in social science, remains ambiguous and attributed this to the fact that many researchers have jumped on to the bandwagon of organizational learning resulting in a multitude of analytical perspectives.

Some scholars view learning as intertwined with organizational routines. It is after all common practice for most organizations to run induction sessions so that employees can learn about standard operational procedures at the workplace. Site inductions, especially in terms of health and safety talks, are another typical example of how routines can be a source of learning in organizations. As Tranfield *et al.* (2000) articulated, these represent the systems approach of learning where the combination of cognitive, structural and behavioural aspects of routines works with resources to create organizational competencies. The transfer of knowledge from the trainer to the trainee, however, is known as "single-loop learning", where learning is absorbed mainly by the recipient of the information. Of course, routines are also continually subjected to the dynamics of change, and can be conceptualised as generative systems for learning (Pentland and Feldman 2005). At times, organizational routines become challenged and adapted, whether explicitly or implicitly through informal practices at the workplace. So, the employee is not merely a passive recipient of knowledge about the routines, but can also actively question the validity and reliability of the information being transferred. Such learning is said to be "double-loop learning". Meetings and away days that encourage employee participation are practical examples of where double-loop learning might take place.

Researchers have also acknowledged that organizations have to create a climate where learning is encouraged. Tangibly, Lipshitz et al. (2002) proffered five key facets for consideration, including the:

- Structural facet: are there mechanisms in place that integrate learning as a critical component of working?
- Cultural facet: are aspects such as transparency, accountability and inquiry encouraged in the working environment?
- Contextual facet: is learning simply an abstract concept or inextricably linked to the task at hand; and is there leadership commitment to learning?
- Psychological facet: does everyone feel safe enough to learn from mistakes made and is there commitment by everyone to learn?
- Policy facet: how is the commitment to learn encouraged by organizational policies, especially in terms of rewards and incentives?

Apart from focusing on procedural elements of learning, other scholars have concentrated on the socialisation aspects of learning. After all, people do play a critical role in the enactment of learning. Wenger (2000), in particular, promoted the notion of communities of practice in the pursuit of learning. Industry network meetings to disseminate regulatory changes or new production methods are examples where communities of practice gather together in the pursuit of learning. Chan et al. (2005) argued that the essence of learning should be brought more to the fore, and that communities of practice should genuinely become communities for learning if progress and development were to be attained. Responding to change merely leads to incremental improvements in the industry and downplays the potential that learning can bring to a step change in the industry. Yet, as Weick (2002) suggested in the "heat of the battle", much organizational learning literature has focused mainly on change as the key driver for learning. Harrison and Leitch (2000), however, were quick to point out that change alone does not imply that learning would take place and hence, change should not be the sole precursor for organizational learning. Harrison and Leitch (2000) suggested that the process of becoming a learning organization was more crucial than the study of the learning organization as a being (i.e. an end) and sought, in their action research case study, to involve individual persons within their case study organizations in "analytical dialogue … as a starting point for a process of self-development and self-awareness" (p. 115). Yet, it is the lack of such critical reflexivity in organizations that can often impede learning and development to take place.

Drawing on Mayhew (2003), Chan and Cooper (2006) suggested that construction companies engaging in business-as-usual can often be blinded to the notion of latent skills shortages, where there are skills gaps that go unrecognised by companies simply because they have coped hitherto without the necessary skills. Drawing on case study research, Chan and Cooper (2006) suggested that latent skills shortages can manifest at three levels (see Figure 6.1). Accordingly, it is crucial for businesses to formulate their long-term strategy and connect this with recruitment and selection, deployment (or resource allocation) and development practices. It would also be ideal if allocation of resources is done on the basis of some form of job analysis, i.e. fitting round pegs in round holes. Similarly, the development needs of employees should be periodically identified through e.g. the appraisal process. It

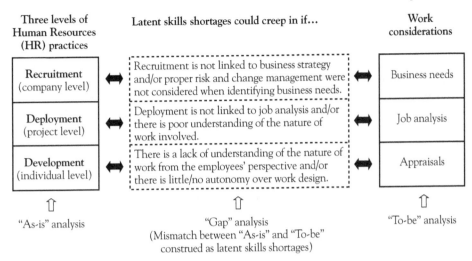

Three levels of Human Resources (HR) practices	Latent skills shortages could creep in if...	Work considerations
Recruitment (company level)	Recruitment is not linked to business strategy and/or proper risk and change management were not considered when identifying business needs.	Business needs
Deployment (project level)	Deployment is not linked to job analysis and/or there is poor understanding of the nature of work involved.	Job analysis
Development (individual level)	There is a lack of understanding of the nature of work from the employees' perspective and/or there is little/no autonomy over work design.	Appraisals

"As-is" analysis

"Gap" analysis
(Mismatch between "As-is" and "To-be"
construed as latent skills shortages)

"To-be" analysis

Figure 6.1 Risk of development of latent skills shortages.

must be added that prescription of a one-size-fits-all model is deliberately avoided here, since it has been acknowledged earlier that the diverse nature of organizations that exist in the construction industry makes a universal model less meaningful. Nonetheless, for companies to internalise HRD practices and to harness the power of organizational learning, it is vitally important that organizations develop a degree of critical reflexivity when aligning their HR practices with strategic considerations at firm, project and individual levels.

Yet Chan and Cooper (2006) found that construction companies rarely think about strategy in a systematic way; instead, there is a tendency for construction companies, driven by the project delivery imperative, to behave tactically. So, recruitment and selection, and allocation of resources, are more likely to be done through quick-fixes since there is often very little time allowed to undertake such activities. As Loosemore *et al.* (2003) commented "projects tend to be won at short notice, requiring the rapid mobilisation of teams to distant locations. In the rush to respond to this situation, individual employee needs can be excluded from the resourcing decision-making process" (p. 84). Perhaps unsurprisingly, Druker and White (1996) observed a wealth of anecdotal evidence that points to the widespread use of informal HRM at all levels in construction, where personal introductions and contacts matter in the selection of staff. Unless construction firms become more reflective in strategically accounting for the development of its human resources, and aspire to become learning organizations, firms will continue to suffer the impacts of latent skills shortages and this reactive *modus operandi* is likely to continue to stymie the long-range thinking necessary to embed development as an integral aspect of organizational development.

Externalising HRD: how might companies engage with external stakeholders

There is a move towards increasing employer engagement in skills development activities. As mentioned earlier, contemporary HRD extends beyond traditional education and training courses formally delivered in a school/college setting. Whilst formal programmes still constitute a large part of development activities, the focus is shifting more towards work-based learning and individual self-development initiatives. This demands greater attention of companies to look outside the firm for their HRD solutions. As was discussed earlier, before employers can decide on the nature of HRD practices, some clarity across the company regarding its strategic direction would be necessary. After all, it does make sense for companies to reflect on the HRD needs and how this aligns with how the company, balancing the developmental needs of its employees and the wider shifts in the marketplace, develops in its business operations. It would also be useful for employers to consider the range of education, training and learning opportunities, and this would involve seeking external advice, where appropriate, on information regarding the availability of education and training courses, grant funding opportunities, as well as consideration of work organization issues to free up time and space for employees to engage in self-learning opportunities.

Another way in which companies, particularly SMEs, can engage in HRD effectively is through inter-firm coordination of HRD. Here, Gospel and Foreman (2006) usefully describe a number of alternatives for consideration, including coordination through:

- chamber of commerce (or equivalent umbrella organization of businesses): this is where the chamber devises training plans and acts as liaison body between employers and providers of education and training. Chambers of commerce tend to have close working relationships with businesses and so coordination of skills development by the chamber can be especially helpful for SMEs who might not have the resources to go through the bureaucracy attached to training;
- industry-wide employers' association: often, these associations offer specific training towards a set of industry-wide, national standards. Members also benefit from the opportunity to recruit and select from a trusted pool of trainees;
- local group training association: this is where companies, particularly SMEs, pool resources together to invest in education and training for their workers, and;
- supply networks: this is usually mobilised in the automotive industry, where training is negotiated across the supply chain and there are opportunities for secondments of personnel across the supply network.

A contemporary example of where principles of inter-firm coordination of skills development are beginning to reinvigorate HRD practices in construction. For example, the MACE Group has set up the MACE Business School to provide continuous professional development courses to its supply chain partners, in part to engender upskilling of its supply chain workforce and to update its supply chain participants on contemporary thinking. Furthermore, this allows the MACE Group to ensure that its supply chain participants are enculturated into the MACE way of working (see www.macebusinessschool.co.uk).

What is the future for HRD in construction?

Thus far, this chapter has emphasised the importance of HRD and learning to the modern construction business. It has also revealed the challenges inherent in construction firms embracing the learning paradigm. Nonetheless, there is a clear need for the industry to shift its thinking away from traditional notions of training towards greater emphasis on reflective learning and empowering individuals to engage in HRD if it were to ensure the continued growth and development of its organizations in the future. The emphasis on learning is likely to continue to grow, and even prevail, in the future of HRD in construction. In order to support organizations in their pursuit of becoming learning organizations, there remains a number of questions and new directions that are worthy of further research which are explored in this section.

Moving towards an interorganizational perspective

The high commitment management (HCM) concept is currently framing much of the debate around the HRM-organizational performance link. Marchington and Wilkinson (2005) adapt Pfeffer's (1998) description in suggesting practices such as employee security, selective hiring and sophisticated selection, extensive training, learning and development, employee involvement, self-managed teams, performance contingent reward and reduction of staff differentials as HR practices consistent with the high commitment ethos. Training, learning and development are *writ large* within this framework as the key source of sustained competitive advantage. However, the nature of HR policies which support commitment remain unclear and would undoubtedly need to be tailored to support the complex, temporary multiple organizational climate which characterises construction firms. Disentangling the ambiguities (see Marchington *et al.* 2009) of who holds the responsibility for training and indeed identifying what aspects of skills and knowledge to develop remain a pressing research and practice issue to resolve. Within construction, the interconnectedness of organizations and the inherent need for interorganizational learning to precipitate positive project outcomes demands an approach which supports learning and development across organizational boundaries and supply chains. Construction offers a problematic context within which to enact 'best practice' HRM in this regard, and provides a practice context which challenges the applicability of much mainstream theory. Examining ways in which learning can be supported both within and between construction organizations is therefore urgently required.

The 'practice-turn' of HRD

Given the questions which exist around 'best practice' HRM and the applicability of prescriptive approaches towards HRD, efforts are required which examine the nature of various approaches to learning and development and their impact on ensuring the sustained dynamic capability of construction firms. Boxall and Macky (2009) suggest that the pathway to better research in high performance work systems lies in developing a better understanding of the experiences of workers and the links to operational outcomes. Exploring the bundles of practices which can be combined to yield particular outcomes within specific operating environments demands the adoption of 'practice perspectives' on HRD to inform the development of existing theoretical models. Practice-based research would therefore seem like an essential prerequisite to informing developments in the application of HCM within the construction context. Perhaps the opportunity for managers is to deliberately design their learning strategy around the externalised view of HRD which sees capability as internally generated but responsive to (and reliant upon) exogenous influences.

Connecting to an individual-collective learning agenda

A key theme undergirding this chapter has concerned the shift in orientation away from 'top-down' HRD models to 'bottom-up' approaches where individual employees take responsibility for their own learning development in a way which is consistent with organizational goals. Whilst this unitary perspective might seem somewhat naïve given the pluralist nature of many organizations, the constantly changing requirements of construction work and the need to continually reconfigure scarce intellectual and knowledge resources to meet customer and project needs necessitate a more organic and responsive approach to HRD than has traditionally been the case. However, moving from a functional development model (bureaucratic, task-oriented) to a personalised dynamic development-based model represents a considerable challenge for construction firms rooted in Taylorist approaches to work organization. The challenge will be in developing the learning climate to sustain such a change (cf. Sloman 2003).

Towards a broader performance view of HRD

The intention of much work within the HRM field has sought to relate HR practice to the performance of the firm. HRD is no exception, with many perspectives rooted in a desire to align employee contributions and behaviour with the strategic intentions of the firm. However, HRD provides an excellent vehicle for construction organizations to move beyond purely commercial objectives to responding to employee needs. As Sambrook (2004) suggests, HRD has a potential humanistic and emancipatory role in helping individuals to meet their own aspirations. In this respect, corporate social responsibility (CSR) drivers for HRD can be seen as increasingly important as part of efforts to reflect the social responsibility to employees' wider growth and development within construction organizations.

Concluding remarks

HRD should be beneficial for individual workers, employers and society at large. However, as has been revealed, the ability of construction firms to embrace HRD practices is impeded by the industry structure within which they operate. The plurality of organizations and corresponding interests that exist in the industry result in divergent conceptualisations of skills which presents a problematic landscape within which to enact enlightened HRD practices.

In this chapter we have deliberately adopted a critical perspective in attempting to understand the HRD function and the reasons as to why construction organizations struggle to enact it effectively. We have also avoided exhortations of 'how to' do HRD within the industry given the heterogeneity of construction organizations and the need for them to respond to a continually changing set of external influences. Nevertheless, a higher priority for HRD is an essential prerequisite of employee well-being and a positive employment relationship, and it is from this that continually improving project and firm-level performance will flow. Working towards the adoption of 'Learning Organization' principles would seem like a potential route for overcoming some of these constraints in a way which is likely to provide construction organizations with the agility to respond to change in a way which is likely to encourage their sustained competitiveness in the future.

Addressing HRD in the contemporary construction organization

- Learning and development are crucially important as a source of competitive advantage, and yet it is surprisingly underutilised as a performative lever within many construction organizations.
- Addressing learning and HRD requires that organizations propagate a learning climate. Within construction, where so much work is subcontracted and outsourced, this

demands an interorganizational perspective through which the temporality of project alliances can be addressed.

- HRD should not be seen as a purely top-down (managerially driven) process, but also one which should be bottom-up (employee pulled). This will inevitably mean having to respond to more individualised learning requirements.
- Finally, it is important to stress that there is no one best way to do HRD in construction. The approach taken must be sensitive to the specific organizational context within which the strategy is practiced. Moreover, the approach should be flexible enough to respond to competitive pressures and to the corresponding changing capability requirements of the organization.

References

Armstrong, M. (2003) *A Handbook of Human Resource Management Practice*, ninth edition, London: Kogan Page.

Barney, J. (1991) Firm resources and sustained competitive advantage, *Journal of Management*, 17(1): 99–120.

Beaney, W.D. (2006) *An Exploration of Sectoral Mobility in the UK Labour Force: A Principle Components Analysis*, unpublished MSc thesis, Newcastle Business School, Northumbria University, UK.

Becker, G.S. (1964) *Human Capital: A Theoretical and Empirical Analysis, with Special Reference to Education*, third edition, Chicago: University of Chicago Press.

Bellini, E. and Canonico, P. (2008) Knowing communities in project driven organizations: Analysing the strategic impact of socially constructed HRM practices, *International Journal of Project Management*, 26(1): 44–50.

BERR (Business Enterprise and Regulatory Reform) Department (2007) *Construction Statistics Annual 2007*, London: BERR

Boxall, P. and Macky, K. (2009) Research and theory on high-performance work systems: progressing the high-involvement stream, *Human Resource Management Journal*, 19(1): 3–23.

Brady, T. and Davies, A. (2004) Building project capabilities: from exploratory to exploitative learning, *Organization Studies*, 25(9): 1601–1621.

Braverman, H. (1974) *Labour and Monopoly Capital: The Degradation of Work in the Twentieth Century*, New York: Monthly Review Press.

Burnes, B., Cooper, C. and West, P. (2003) Organisational learning: the new management paradigm? *Management decision*, 41(5): 452–464.

Castells, M. (2002) *Internet and the network enterprise*, plenary address given to the 18th EGOS (European Group for Organizational Studies) Colloquium, Barcelona, July 4.

Chan, P. (2007) Managing human resources without Human Resources Management Departments: some exploratory findings on a construction project, in W. Hughes (ed.), *Proceedings of the CME25: Construction Management and Economics: Past, Present and Future*, 16–18 July 2007, University of Reading, UK.

—— and Cooper, R. (2006) Talent management in construction project organisations: do you know where your experts are? *Construction Information Quarterly*, 8(1): 12–18.

—— and Kaka, A. (2007) The impacts of workforce integration on productivity, in: A.R.J. Dainty, S. Green and B. Bagilhole (eds), *People and Culture in Construction: A Reader*, London: Spon, pp. 240–257.

Chan, P.W., Cooper, R. and Tzortzopoulos, P. (2005) Organisational learning: conceptual challenges from a project perspective, *Construction Management and Economics*, 23(7): 747–756.

Clarke, L. (1992) *The Building Labour Process: Problems of Skills, Training and Employment in the British Construction Industry in the 1980s*, Occasional Paper No. 50, Englemere: CIOB.

—— and Herrmann, G. (2004) Cost vs. production: disparities in social housing construction in Britain and Germany, *Construction Management and Economics*, 22(5): 521–532.

Cockburn, C. (1983) *Brothers: Male Dominance and Technological Change*. London: Pluto Press.

Dainty, A.R.J., Green, S.D. and Bagilhole, B.M. (2007) People and culture in construction: contexts and challenges, in A.R.J. Dainty, S.D Green and B.M. Bagilhole (eds) *People and Culture in Construction: A Reader*, Abingdon: Taylor and Francis, pp. 3–25.

Dainty, A.R.J., Ison, S.G. and Root, D.S. (2005) Averting the construction skills crisis: a regional approach, *Local Economy*, 20(1): 79–89.

Dainty, A.R.J., Raidén, A.B. and Neale, R.H. (2009) Incorporating employee resourcing requirements into deployment decision-making, *Project Management Journal*, 40(2): 7–18.

Davies, A. and Brady, T. (2000) Organisational capabilities and learning in complex product systems: towards repeatable solutions, *Research Policy*, 29(7–8), 931–953.

Davis, P.R. and Walker, D.H.T. (2009) Building capability in construction projects: a relationship-based approach, *Engineering, Construction and Architectural Management*, 16(5): 475–489.

De Haan, J., Voordijk, H. and Joosten, G.J. (2002) Market strategies and core capabilities in the building industry, *Construction Management and Economics*, 20(2): 109–118.

Druker, J. and White, G. (1995) Misunderstood and undervalued? Personnel management in construction, *Human Resource Management Journal*, 5(3): 77–91.

—— and —— (1996) *Managing People in Construction*, London: Institute of Personnel and Development.

Eisenhardt, K.M. and Martin, J.A. (2000) Dynamic capabilities: what are they? *Strategic Management Journal*, 21: 1105–1121.

Forde, C. and MacKenzie, R. (2007) Getting the mix right? The use of labour contract alternatives in UK construction, *Personnel Review*, 36(4): 549–563.

Garavan, T.N., Heraty, N. and Barnicle, B. (1999), Human resource development literature: current issues, priorities and dilemmas, *Journal of European Industrial Training*, 23(4): 169–179.

Gospel, H. and Foreman, J. (2006) Inter-firm training coordination in Britain, *British Journal of Industrial Relations*, 44(2): 191–214.

Grant, R. (1996) Towards a knowledge-based theory of the firm, *Strategic Management Journal*, 17: 109–122.

Green, S.D., Larsen, G. and Kao, C.C. (2008) Competitive strategy revisited: contested concepts and dynamic capabilities, *Construction Management and Economics*, 26(1): 63–78.

Grimshaw, D. and Rubery, J. (2005) Inter-capital relations and the network organisation: redefining the work and employment nexus, *Cambridge Journal of Economics*, 29(6): 1027–1051.

Groák, S. (1994) Is construction an industry? Notes towards a greater analytic emphasis on external linkages, *Construction Management and Economics*, 129(4): 287–293.

Groen, J.A. (2006) Occupation-specific human capital and local labour markets, *Oxford Economic Papers*, 58(4): 722–741.

Grugulis, I. (2003a) Putting skills to work: learning and employment at the start of the century, *Human Resource Management Journal*, 13(2): 3–12.

—— (2003b) The contribution of National Vocational Qualifications to the growth of skills in the UK, *British Journal of Industrial Relations*, 41(3): 457–475.

——, Vincent, S. and Hebson, G. (2003) The rise of the 'network organisation' and the decline of discretion, *Human Resource Management Journal*, 13(2): 45–59.

——, Warhurst, C. and Keep, E. (2004) What's happening to 'skill'? in C. Warhurst, E. Keep and I. Grugulis (eds), *The Skills that Matter*, Basingstoke: Palgrave Macmillan, pp. 1–19.

Harrison, R.T. and Leitch, C.M. (2000) Learning and organization in the knowledge-based information economy: initial findings from a participatory action research case study, *British Journal of Management*, 11(2): 103–119.

Hoskisson, R.E., Hitt, M.A., Wan, W.P. and Yiu, D. (1999) Theory and research in strategic management: swings of a pendulum, *Journal of Management*, 25(3): 417–456.

Huselid, M.A. (1995) The impact of human resource management practices on turnover, productivity, and corporate financial performance, *Academy of Management Journal*, 38(3): 635–672.

Huysman, M. (2000) An organisational learning approach to the learning organisation, *European Journal of Work and Organizational Psychology*, 9(2): 133–145.

Kinnie, N.J., Swart, J. and Purcell, J. (2005) Influences on the choice of HR system: the network organisation perspective, *International Journal of Human Resource Management*, 16(6): 1004–1028.

Kululanga, G.K., Edum-Fotwe, F.T. and McCaffer, R. (2001) Measuring construction contractors' organisational learning, *Building Research and Information*, 29(1): 21–29.

Lähteenmäki, S., Toivonen, J. and Mattila, M. (2001) Critical aspects of organisational learning research and proposals for its measurement, *British Journal of Management*, 12(2): 113–129.

Lipshitz, R., Popper, M. and Friedman, V.J. (2002) A multifacet model of organisational learning, *Journal of Applied Behavioural Science*, 38(1): 78–98.

Loosemore, M., Dainty, A. and Lingard, H. (2003) *Human Resource Management in Construction Projects: Strategic and Operational Approaches*, London: Spon Press.

Mackenzie, S., Kilpatrick, A.R. and Akintoye, A. (2000) UK construction skills shortage response strategies and an analysis of industry perceptions, *Construction Management and Economics*, 18(7): 853–862.

Marchington, M. and Wilkinson, A. (2005) *Human Resource Management at Work*, third edition, London: Chartered Institute of Personnel and Development.

——and —— (2008) *Human Resource Management at Work: People Management and Development*, fourth edition, London: Chartered Institute of Personnel and Development.

——, Carroll, M., Grimshaw, D., Pass, S. and Rubery, J. (2009) *Managing People in Networked Organisations*, London: Chartered Institute of Personnel and Development.

——, Cooke, F.L. and Hebson, G. (2010) Human resource management across organizational boundaries, in: A. Wilkinson, N. Bacon, T. Redman and S. Snell (eds), *The SAGE Handbook of Human Resource Management*, London: Sage, pp. 460–474.

Mayhew, K. (2003) *The UK Skills and Productivity Gap*, talk given to the CIHE/AIM colloquia, 30 October.

Murray, M.D. and Dainty, A.R.J. (2009) *Corporate Social Responsibility in the Construction Industry*, Abingdon: Taylor and Francis.

Nyhan, B., Cressey, P., Tomassini, M., Kelleher, M. and Poell, R. (2004) European perspectives on the learning organisation, *Journal of European Industrial Training*, 28(1): 67–92.

Pearce, D. (2003) *The Social and Economic Value of Construction: The Construction Industry's Contribution to Sustainable Development*, London: nCRISP.

—— (2006) Is the construction sector sustainable? Definitions and reflections, *Building Research and Information*, 34(3): 201–207.

Pedler, M., Burgoyne, J. and Boydell, T. (1991) *The Learning Company*, London: McGraw-Hill.

Pentland, B.T. and Feldman, M.S. (2005) Organizational routines as a unit of analysis, *Industrial and Corporate Change*, 14(5): 793–815.

Pfeffer, J. (1998) *The Human Equation: Building Profits by Putting People First*, Boston, MA: Harvard Business School Press.

Purcell, J. (1999) The search for 'best practice' or 'best fit': chimera or cul-de-sac? *Human Resource Management Journal*, 9(3): 23–41.

Raidén, A.B. and Dainty, A.R.J. (2006) Human resource development in construction organisations: an example of a 'chaordic' learning organisation? *The Learning Organisation*, 13(1): 63–79.

Sambrook, S (2004) A 'critical' time for HRD? *Journal of European Industrial Training*, 28(8–9): 611–624.

Senge, P.M. (1990) *The Fifth Discipline: The Art and Practice of the Learning Organization*, New York: Doubleday Dell.

Simmonds, D. and Pedersen, C. (2006) HRD: the shapes and things to come, *Journal of Workplace Learning*, 18(2): 122–134.

Sloman, M. (2003) *Training in the Age of the Learner*, London: Chartered Institute of Personnel and Development.

—— (2005) *Training to Learning: Change Agenda*, London: Chartered Institute of Personnel and Development.

Stata, R. (1989) Organisational learning: the key to management innovation, *Sloan Management Review*, 30(3): 63–74.

Teece, D.J., Pisano, G. and Shuen, A. (1997) Dynamic capabilities and strategic management, *Strategic Management Journal*, 18(7): 509–533.

Tranfield, D., Duberley, J., Smith S., Musson, G. and Stokes, P. (2000) Organisational learning: it's just routine! *Management Decision*, 38(4): 253–260.

Weick, K.E. (2002) Puzzles in organisational learning: an exercise in disciplined imagination, *British Journal of Management*, 13(Supp. 2): S7–S15.

Wenger, E. (2000) Communities of practice and social learning systems, *Organisation*, **7**(2): 225–246.

Wernerfelt, B. (1984) A resource-based view of the firm, *Strategic Management Journal*, 5(2): 171–180.

Winch, G. (1998) The growth of self-employment in British construction, *Construction Management and Economics*, 16(5): 531–542.

Wright, P.M., Dunford, B.B. and Snell, S.A. (2005) Human resources and the resource based view of the firm, in G. Salaman, J. Storey and J. Bilsbery (eds), *Strategic Human Resource Management Theory and Practice: A Reader*, London: Sage, pp17–39.

7 Globalization

S. Ping Ho

Introduction

Although the construction industry is 'local' by its nature in terms of many factors such as regulatory, political, social, and procurement conditions (Ofori 2003), the globalization of construction is becoming an inevitable reality under today's globalization movement. From the supply side, more firms will join the global market because the construction industry shares similar traits with other industries that facilitate globalization, such as the reduced costs of communication, transportation and funds. From the demand side, the breakdown of trade barriers has created more opportunities for firms with competitive advantages. For example, as more multinational enterprises (MNEs) move out of their domestic markets because of the growth of multinational operations, these MNEs may demand their suppliers, including constructors, to continue working with them in foreign markets (Ngowi *et al.* 2005). As a result of global competition, staying domestic is no longer safe from competition and, maybe, the best strategy for firms to survive the increasingly intensive competition is to grow and become internationals. In the past decade, we have observed several important developments and trends in international construction as predicted by various scholars. For example, according to Ofori (2003), many scholars predicted that the share of demand for large projects would increasingly move towards Asia-Pacific, Africa, and South America. Among the countries in these areas, China, Vietnam, Malaysia, India and Russia would have the fastest growing construction markets (Bon and Crosthwaite 2001). We have also observed that the number of international firms from middle-income and developing countries has been increasing. As such, there are increasing concerns and studies on the emerging forces/MNEs from countries such as China and Korea and on the emerging markets such as China, India, and Vietnam. However, as predicted by Ofori (2003), construction firms from Western Europe and North America would continue to have competitive advantage in highly specialized services.

Extant research on construction internationalization/globalization has been focusing on issues such as the globalization impacts and trends, the key successful factors and risks in global projects and internationalization process, the emerging international markets or contractors, and the use of joint ventures (JVs). However, as the global trend becomes certain but the environment becomes more complex, as noted by Levitt (2007) in discussing the construction, engineering and management research for the next 50 years, the new challenges of construction globalization will be centered on the integration of non-traditional concerns such as culture, institutions, organizations, social mechanisms, alternative project deliveries and financing, and sustainability, etc. In this chapter, we shall adopt some of the above non-traditional views and examine three important issues concerning the organization strategies for competing in the move of construction globalization.

The first issue to be addressed is the internationalization process of construction firms. We will begin by introducing two important theories in internationalization that try to answer why and how firms conduct their business in foreign countries, whereas firms will incur significant additional costs associated with trading internationally. Then we will present a framework for the internationalization process of construction firms based on Ho (2009) and derive implications for internationalization strategies, including the strategies for capability building, market locations, and entry modes.

The second issue focuses on the management of joint ventures (JVs) in construction. In the construction industry, JVs have become one of the major organizational forms utilized in global projects. Among various management challenges, in this chapter, we shall introduce a model based on Ho et al. (2009b, 2009c) and examine some theories and the evidence on how to manage construction joint ventures (CJVs) through better choices of governance structures.

The third issue concerns how to select collaborating partners with respect to different contingency factors. Because of the effectiveness of using JVs/alliances to penetrate a market or perform global projects, partner selection has become an important issue for firms that are involved in global projects. However, since "good" partners are scarce, most partner candidates can only meet some of the criteria and waiting for partners with "perfect" fit will be too costly to be justified. Therefore, it is essential to differentiate the relative importance of partner selection criteria. In this chapter, we will introduce a game theory based model by Ho et al. (2009a) that analyzes the partner selection problems and suggests strategies under multiple contingency contexts.

The internationalization process of construction firms

Dunning's OLI paradigm, or the eclectic theory of international production (Dunning 1988), is one of the most important theories in internationalization. The OLI paradigm integrates three different FDI (Foreign Direct Investment) theories, each with a different focus or question. FDI can be defined as a company from one country making a physical investment that produces merchants directly in another country. Broadly defined, FDI can be considered the establishment of a business entity or enterprise, such as a subsidiary or branch office, that performs one of the major functions of its parent company. FDI can also be regarded as an entry mode that demands the highest resource commitment compared to other modes such as export or franchise. One objective of FDI is to answer why firms should conduct their business in other countries given the fact that firms will incur significant additional costs due to resource mobilization, the lack of local knowledge and supporting resources, and various risks associated with international trade. Whereas the transaction cost economics (TCE) view explains the boundary of firms in terms whether or not they are using "hierarchy" to internalize certain transaction activities, TCE's explanation of FDI is not satisfactory. Dunning added two more factors other than TCE-based I (Internalization advantages), namely O (Ownership advantages) and L (Location advantages), to address the why, where, and how questions of FDI altogether through O, L, and I, respectively.

The ownership advantages view of FDI suggests that the MNE must possess one or more firm specific advantage, such as core competence or economies of scale, which allow it to overcome the additional costs of operating in a foreign country. The location advantages view of FDI emphasizes that the factor advantages, such as cheap labor or market potential, in a foreign country may further strengthen the firm's competitive advantages in earning rents and motivates the MNE to extend its operation to a particular foreign country.

The Uppsala model or U-model, proposed by Johanson (faculty of the University of Uppsala) and Wiedersheim-Paul (1975) and Johanson and Vahlne (1977), is another important internationalization model. The U-model is a process model, explaining the typical incremental nature of the internationalization process. The most important insight of the model is that the resource commitment to a foreign market is increased *incrementally* because the knowledge of the foreign market can only be *gradually* learned during the process. As more market knowledge is learned, firms can then proceed to the stages or use the entry modes that represent higher degrees of international involvement or resource commitment. From the learning and process perspectives, the U-model can be considered a dynamic model, as opposed to the relatively static OLI paradigm. In a dynamic model, the internationalization strategy such as entry mode or the selection of foreign markets/countries will change as the firms move to a different state/stage of internationalization. However, the U-model has two major constraints. First, the model does not explain why or how the process starts (Anderson 1993). Second, market knowledge seems to be narrowly considered as the only or major factor in determining the resource commitment decision.

A dynamic OLI (D-OLI) model for construction internationalization

The pre-international stage of internationalization process

In the global economy, as trade barriers are removed at an increasing speed, MNEs enter local markets as new competitors and, at the same time, the global market provides vast business potential for firms with competitive advantages. As a result, firms with ownership advantages (O), in conjunction with location advantages (L), are motivated to start their internationalization process by both "push" and "pull" market forces. In the past, MNEs in construction were mainly "pushed" out from the advanced industrialized countries because market demands for construction in these countries could no longer support those large and capable firms. As a result, these firms with advanced technology, financing capability, and the ability to manage large, complex projects moved to foreign markets where their ownership advantages can be leveraged. Nevertheless, new market opportunities and reduced costs due to modern globalization have further "pulled" firms with competitive advantages out to international markets. This is evident by the fact that the number of international construction firms from developing countries has been increasing lately. The ownership advantages and location advantages of these firms include "access to inexpensive, highly skilled labor proficient in advanced technology and/or close geographical, cultural and language proximity to their markets" (Ofori 2003). What is going to happen in the next two to three decades is that more local firms in those once closed markets will be "pushed" and "pulled" out of the their local markets and become the new players in the global market. To a degree, in order to survive today's global competition, construction firms will be driven to learn from their collaborating partners and build reputations by working with reputable partners on noteworthy projects in order to become MNEs or international contractors, or, at least, international-capable contractors.

These international-capable firms are capable of collaborating with MNEs in performing global projects. They may still stay local but collaborate with MNEs as a major partner to perform local projects. They may sometimes go abroad and collaborate with MNEs as a supporting partner to compete for international projects. We shall call these firms "the pre-internationals." As such, modern firms in the global economy have to cultivate those ownership advantages required for global competition, instead of just the market knowledge emphasized in "the U-model". From this perspective, JVs may be a better choice of entry mode during the initiation stage of the internationalization process. Consequently, the market location may be changing frequently and will be a "partner following strategy;" that is, the pre-internationals will follow their major collaborating partners, mostly experienced MNEs, to those countries that allow the firms to leverage their resources and learn. Often, in this stage, the firms will take the role of minority JV partners or even subcontractors. The "partner" followed can also be a firm's previous client from the home country.

The pre-MNE stage of internationalization process

As firms accumulate sufficient capabilities, market knowledge, resources, or reputations required for higher international involvement, they become international-ready and can take the role of project leaders in JVs or even to establish sole venture enterprises or subsidiaries in some countries. We shall call the firms at this stage the pre-MNEs, who begin to be capable of playing a major role in undertaking a global project or establishing a few subsidiaries. In this stage, the pre-MNEs do not yet have full capacity to compete or commit resources globally and their competitiveness can still be constrained by their scale, financing ability, market knowledge, or human capital etc. As such, in terms of market location strategy, the firms may limit their market locations to only a few that allow them to exploit the location and internalization advantages. We shall term this location strategy as "focused market seeking strategy."

As emphasized in the U-model, in this stage, firms will focus on the continued cycles of learning of market knowledge resulting from the increased commitment to foreign markets. Market knowledge becomes critically important because the pre-MNEs tend to use high degrees of international involvement in a few focused foreign markets and it is crucial that the firms learn about those markets as much, as deeply as possible. From the OLI perspective, location advantages (L) can be best exploited in this stage because most market knowledge learned through the focused market-seeking strategy is location specific. The market knowledge accumulated from different countries will also prepare the pre-MNEs to become more globally minded and integrated. Internalization advantages (I) are also very important in this stage because the construction industry typically is highly protected by a country and using a JV with a company or sole venture company is almost the only way to access the market's regular projects that are not yet opened for global competition. As such, the pre-MNE stage suggests the use of the entry modes with higher resource commitment, such as JVs with a majority share, local agent, JV companies, branch office/company, or even a sole venture company.

The MNE stage of internationalization process

The MNE stage is the final stage of the internationalization process. In this stage, the firms are becoming MNEs as they can compete globally and have substantial support from their foreign subsidiaries. As we have observed in the construction industry, the MNEs can undertake projects in almost any country with few concerns about the "psychic distance." Their employees are constantly involved in global projects, familiar with different major types of culture, and prepared for potential collaboration problems and complex environments. The MNEs can efficiently integrate key resource and effectively work with partners and manage key subcontractors in global projects. Very often, the MNEs perform a project by having employees from several international subsidiaries working together as a team via modern communication and networking technology, and knowledge management systems. In this way, the experts of an MNE can provide services worldwide and the MNE can significantly reduce the costs of operating a subsidiary in terms of the fixed investment on human capital.

In this stage, the MNEs focus on integrating and leveraging their OLI advantages to their maximum levels. On the one hand, the MNEs establish a network of subsidiaries that may work together or complement one another in terms of their different location and ownership advantages. On the other hand, the MNEs regard the world as one market and compete for those global projects that promise most profits. As such, in terms of market location strategy, the MNEs in construction will adopt the "global market seeking strategy" so as to compete globally and integrate their global subsidiaries. In terms of entry mode strategy, JVs, branch offices, and sole venture companies will be the dominant modes, given their aforementioned global strategy. It is worth noting that, in this stage, it is not necessarily true that the market knowledge emphasized in the U-model can only be obtained by sequentially increasing the resource commitment. MNEs, based on their strong ownership advantages, may effectively obtain the required market knowledge within a short period of time through the means of JVs or mergers and acquisitions.

Governance structure choices of international construction joint ventures

As illustrated in the previous section, JVs are a popular organizational form often used in global projects and also an important entry mode for MNEs. JVs are critical to large scale, complex projects because complementary resources can be pooled or integrated to complete a project, especially for those whose host countries do not have enough capacity or required technology. However, because of the complexity of JVs, the management of JVs is much more difficult than that of usual projects. Research on construction alliances has focused on issues such as:

1. rationales and benefits behind international construction alliances (Badger and Mulligan 1995; Sillars and Kangari 1997);
2. governance structures of construction alliances (Ngowi 2007; Chen 2005);
3. performance or organizational success in alliances or joint ventures (Luo 2001; Mohamed 2003; Sillars and Kangari 2004; Ozorhon *et al.* 2007; Ozorhon *et al.* 2008).

There is rich literature in the management discipline regarding strategic alliances and governance structure; however, many relevant findings or theories in management literature need to be re-evaluated when they are applied to the construction industry because the characteristics of JVs are very different from that discussed in management literature.

JVs and CJVs in the construction industry

Construction practitioners often use the term "JVs" as a short name for "CJVs," the project-based construction JVs. In this case, the term "JVs" is regarded as project-based "CJVs." However, in general and also in this chapter, JVs is defined broadly, including CJVs, JV companies, and other types of non-equity alliances such as partnering. Using the aforementioned equity and non-equity taxonomy of JVs, JV companies can be considered equity JVs and partnering can be considered non-equity JVs. However, it is not very straightforward to understand why CJVs can also be considered as equity JVs. CJVs, a special arrangement in construction for performing projects, usually refer to the collaboration through written contracts that enable contractors to share money, abilities, and resources in the duration of a single project (Naylor and Lewis 1997; Morris 1998). CJVs differ from the equity JVs in that creating a new legal entity is not necessarily required in a CJV. Nonetheless, in CJVs, the legal and financial bindings between JV partners are no less than that of a new entity. The practice in construction is that, should the project go wrong, all partners are legally responsible for the consequences no matter how tasks are divided among partners or specified in agreements. In this regard, CJVs could provide no less security to project owners than that in typical equity JVs. In typical equity JVs, the partners are only liable up to the limited equity or assets invested in the new organization, but the CJV partners' liabilities can be up to the total equities of their parent companies. The provision of high security to project owners may be one of the major reasons why CJVs are a popular form of governance structure. Therefore, although usually no new entity is legally required, a CJV can be considered an "equity JV" due to a CJV's strong legal and financial bindings between partners toward their responsibilities to the owner.

Governance structures of CJVs: Jointly Managed JVs (JMJ) and Separately Managed JVs (SMJ)

Following Geringer and Hebert (1989) and Mjoen and Tallman's (1997) contributions in JV control and the operationalization of control, the following perspectives will be adopted to characterize and differentiate the control and governance structure of CJVs:

1. the technical and financial responsibilities and claims associated with each partner;
2. the extent to which major decision making is decentralized to partnering firms; and
3. the levels and needs for coordination

Accordingly, two distinctive organizational control structures for CJVs are identified and defined: Jointly Managed JVs (JMJ) and Separately Managed JVs (SMJ). Although the terms "Integrated JVs" and "Non-integrated JVs" are also found in construction practice and literature (Chen 2005) to describe two different modes of governance structures, we prefer not to use these semantically "strong" terms.

JMJ is characterized by (1) all partners jointly sharing profits and risks of a CJV according to an agreed proportion even though distinctive tasks may still be assigned to each firm; (2) the CJV management team making major decisions, which will be followed by all partners; and (3) the need for coordination and communication being extended to all levels of a CJV organization. On the other hand, SMJ is characterized by (1) each firm being technically and financially responsible for its assigned tasks, which are often negotiated; (2) each firm making most decisions related to the assigned tasks without the needs of consent from other CJV partners; and (3) the need for coordination and communication being limited to higher level managers and are minimum for individuals. In practice, the actual governance structure of a CJV should be somewhere in the spectrum between the two extremes. Thus, here the concern of choosing between JMJ and SMJ will be more precisely regarded as the finer-grained governance structure choice within the spectrum between JMJ and SMJ.

The governance structure model and propositions

Based on Ho *et al.* (2009b), a framework integrating the relational view and resource-based view (RBV) can provide a more comprehensive explanation of the governance structure choices in CJVs. From a relational view, good interfirm traits, also called relational capital, can help to alleviate the interorganizational conflicts and opportunism that can occasion high transaction costs or a premature breakdown of the relationship (Zaheer *et al.* 1998). Relational capital such as trust or reciprocal commitments also helps to extend critical resources beyond firm boundaries to become interorganizational competitive advantages (Dyer and Singh 1998).

In addition, Ho *et al.* incorporate the resource-based view of organizational control, which emphasizes value creation and sustainability of competitive advantages of a firm through the continuous accumulation and utilization of valuable tangible or intangible resources (Das and Teng 2000; Wernerfelt 1984). The exploitation of complementary resources and learning from partners through JVs are considered one of the major sources that lead to a long-lived competitive advantage. Following the logic of RBV, different firms should take into account the characteristics of their specific resources and advantages to pursue different strategies for profits.

Based on the perspective that integrates the relational view and RBV, Ho et al. developed a governance structure decision model for CJVs, proposing four propositions, introduced below, that specify the relationships between the main determinants and the CJV governance structure choices.

Proposition 1: *A CJV with larger cultural difference among partners is more likely to adopt SMJ, while a CJV with smaller cultural differences is more likely to adopt JMJ.*

Organizational culture refers to the set of values, beliefs, understandings, and ways of thinking that are common to the members of an organization (Daft 2001). Many problems experienced by firms in JVs can be traced back to cultural difference (Meschi 1997; Horii *et al.* 2004). Greater cultural distance often results in greater differences in their organizational and administrative practices, employee expectations, and interpretation of and response to strategic issues (Park and Ungson 1997). These differences are often the source of conflicts and detrimental to the relation and mutual trust between partners during the performance of a project. Corporate cultural difference plays an important part in making the choice of governance structure because it often increases the transaction costs, such as monitoring and coordination costs. The impacts of corporate cultural difference are especially emphasized in CJVs because problems resulted from corporate cultural difference are more difficult to resolve in CJVs due to limited project duration. Therefore, through limited interdependence between partners, the SMJ structure helps to reduce the culture-associated frictions and increase the stability of a CJV. On the other hand, a low level of cultural difference provides an environment for working closely and smoothly, which is consistent with the characteristics of JMJ.

Proposition 2: *A CJV with greater trust among partners is more likely to adopt JMJ, while partners with less trust among them will tend to adopt SMJ.*

Trust is built upon an expectation that one partner has for another in the partnership such that their interaction is predictable and the behavior and responses are mutually acceptable to one another (Harrigan 1985). Trust among firms indicates the positive belief that a partner will not take advantage of other partners (Powell 1990). Since close collaboration and interdependence in JMJ usually involve asset-specific investment, in the form of capital and/or labor, and thus, will expose the party who contributes more or first to a high risk of being exploited. As a result, the equilibrium of the collaboration game will be to act opportunistically or conservatively if there is no sufficient trust. Therefore, one can infer that JMJ can achieve its desired level of performance only when sufficient trust between partners exists. On the other hand, when there is no sufficient trust, the use of the SMJ structure may help to reduce the potential conflicts and instability problem by better division of responsible tasks and risks, but can still achieve the goal of pooling complementary resources, although the performance of using SMJ may tend to be less than that in JMJ with trust.

Proposition 3: *A CJV where partners have higher needs for procurement autonomy is more likely to adopt SMJ, while a CJV with fewer needs from partners for procurement autonomy is more likely to adopt JMJ.*

In the construction industry, the success of a contracting firm lies heavily in its capability to acquire inputs at the best price, quality, and reliability (Warszawski, 1996). Here the procurement autonomy is defined as the decisional power to select a firm's preferred suppliers or subcontractors for a JV project. According to the RBV, the procurement advantage may help to obtain favorable tangible resources, such as cheaper materials or capable subcontractors. Depending on the conditions of a JV, sometimes the procurement autonomy will become an important factor that affects a firm's profitability in a JV. For example, when there is serious information asymmetry in the market, procurement through specific channels that are more informed or trusted may reduce transaction costs significantly. They may center on higher flexibility in choosing their own subcontractors or suppliers, and demand more independence between partners. This can be understood from the "focus" dimension of control argued previously. That is, when profits from a procurement advantage are strategically important to a partnering firm it is desirable to balance the control through various procurement arrangements, where each partner has focused control over specific procurement scope. For example, procurements can be decentralized and divided according to dollar amount, assigned tasks, or each partner's comparative purchase advantage. Thus, the needs for procurement autonomy may become a motivating factor for adopting the SMJ structure in a JV to ensure the maximum procurement autonomy.

Proposition 4: *A CJV with stronger motivation in partners for learning is more likely to adopt JMJ, while a CJV where partners are less motivated in learning is more likely to adopt SMJ.*

A firm's organizational learning capability can create competitive advantages (Ulrich and Lake 1991; Inkpen and Crossan 1995). Learning helps to achieve the objective of internalizing the desired external intangible resources such as know-how and expertise. In many cases, learning from partners is one of the major reasons that a firm participates in a CJV.

A JV firm that has a strong motivation for learning will try to exert particular controls or influence over the organization that may facilitate the internalization of its partner's know-how for private gains. In terms of governance structure, JMJ may provide a better environment for learning because it provides a unified chain of command, requires more cooperation between individuals from different partners, and integrates operating procedures. In particular, since "learning by doing" is a major approach to obtaining new knowledge and skills in the construction industry, learning is often naturally achieved under joint operation and management even though the partner with advanced knowledge may not intend to transfer the knowledge. On the other hand, due to the characteristics of firms' resources, some firms in CJVs may not have the needs to internalize other partners' knowledge. For instance, firms may participate in CJVs primarily for entering a new or unfamiliar market, instead of internalizing particular technology or complementary resources. Therefore, from the organizational control perspective, when learning is not an objective of a CJV, partners may prefer SMJ as the control structure.

Partner selection in global projects

In global projects, participants will work with partners from different countries with diverse institutional differences and, thus, often result in very unique challenges and complex interactions, which often lead to significant increased costs (Mahalingam and Levitt 2007) or failures. Therefore, it is not surprising to see that the reported failure rate of such international collaborative activities is high (Khanna *et al.* 1998; Hitt *et al.* 2000). Many potential causes have been studied but the selection of a collaborating partner is particularly one of the major concerns (Geringer 1991; Saxton 1997; Luo 1997; Hitt *et al.* 2000; Sarkar *et al.* 2001; Shah and Swaminathan 2008). The success of any given alliance depends on the extent to which partners match with each other in an effective and efficient manner; that is, the degree of "fit" among alliance members is critical. According to Gale and Luo's (2004) study of international joint ventures in China, "selection of suitable partners" was ranked by both Chinese and foreign partners as the most crucial factor leading to the success of JVs. However, since good partners are scarce (Dyer and Singh 1998), most partner candidates can only meet some of the criteria and seeking or waiting for partners with perfect fit in all dimensions will be costly and difficult to justify economically. Therefore, it is essential to differentiate the relative importance of partner selection criteria.

Following the contingency view, it has been widely recognized that the notion of a good partner is contingent on different contexts. Examples of these contextual variables include the types of markets (Hitt *et al.* 2000; Luo 1997), different institutional environments, and different project characteristics (Shah and Swaminathan 2008), etc. However, most of these studies only examine either one contingency factor or multiple factors individually and fail to consider potential interactive or configurational effects of multiple contexts. Here we shall present a game theory-based framework, capable of analyzing the problems of partner selection in global projects under a configuration of contexts.

Partner fit and partner selection problems

Various dimensions have been used to differentiate partners in literature, among which, the complementary/task-related traits and the compatibility/partner-related traits (Harrigan 1988; Geringer1991; Sarkar *et al.* 2001) are the most often used taxonomy for selection criteria. When partners' traits match the selection criteria (needed traits), it is considered a fit. In literature, the matches of complementary/task-related traits and compatibility/partner-related traits are often regarded as "strategic fit" and "organizational fit," respectively.

A. Strategic fit

Strategic fit is one of the most common and rational explanations for the way in which the strategic and resource needs of alliance partners are met. Based on the resource-based view, strategic fit can be defined as the fit between the firms' traits that represent "what (resources) we have and what (resources) we need." Typical strategic traits may be described by resources of strategic importance such as know-how/knowledge, technologies, market power, market access, unique competencies, and intangible assets such as goodwill, etc. Moreover, firms may elect to form alliances to gain quick access to new geographic or product markets (e.g. Eisenhardt and Schoonhoven 1996). Here it is assumed that strategic fit will center on resource complementarity and be a precondition of partner fit in all scenarios.

B. Organizational fit

Organizational traits affect the efficiency and effectiveness of interfirm cooperation. Here organizational fit is defined as the fit between the firms' traits that represent "who we are and how we perform." Examples of organizational traits include firm size, background, management style, and corporate culture or value system, etc. Bucklin and Sengupta (1993) highlight organizational compatibility as the critical factor for partner selection; that is, not only are the pursued goals shared by partners but also business logic and culture are compatible. In an international alliance, the possible conflicts as a result of economic distance and cultural distance among partners can be effectively reduced by traits such as compatible management styles and corporate cultures, which are collectively characterized as organizational fit. In this model, the organizational fit is further differentiated into mutual fit and unilateral fit given the fact that the fit for one partner does not necessarily lead to the fit for other partners and that it is more difficult and costly to achieve mutual organizational fit.

C. Relational fit

From a relational view, good interfirm traits, also called relational capital, can help to alleviate the interorganizational conflicts and opportunism that can occasion high transaction costs or a premature breakdown of the relationship (Zaheer *et al.* 1998). Relational capital, such as trust or reciprocal commitments, also helps to extend critical resources beyond firm boundaries to become interorganizational competitive advantages (Dyer and Singh 1998). In this model, the relational fit between partners shall be considered the third type of fit. Relational fit is based on the relational view of interfirm cooperation that firms can jointly develop relationships that generate competitive advantages (Dyer and Singh 1998). For example, interpartner trust is characterized as the expectation held by one firm that another will not exploit its vulnerabilities when faced with the opportunity to do so (Barney and Hansen 1994; Mayer *et al.* 1995). This expectation is realized when parties in an alliance demonstrate reliability, act fairly, and exhibit goodwill (Dyer and Chu 2003). Because relational rents or competitive advantages can be better generated under certain conditions when relational capital is presented interorganizationally (Dyer and Singh 1998), relational fit among partners can be considered an objective of an alliance to strive for. Thus, in this model, the relational fit can be further differentiated into "relational fit as a precondition" and "relational fit as a collaboration objective."

Partner selection criteria: firm traits versus types of fits

According to Hitt *et al.* (2000), developed market firms try to leverage their resources through partner selection and emphasize unique competencies, market knowledge/access, and previous alliance experience in their partner selection. This type of study usually will

Table 7.1 Contingency equilibrium solutions

		Complex	Regular	Simple
Local market expansion	F_M: High	$[C_P,C_M]$ in I.	If $L_P<A_P-T_P$: $[U_P,C_M]$ in I. If $L_P>A_P-T_P$: $[C_P,C_M]$ in I.	Same as the left.
Project profit orientation	F_M: Low	$[C_P,C_M]$ in II.	If **Partner dominates:** $[U_P,C_M]$ in IV. If **MNE dominates:** $[U_M,C_P]$ If **no dominant firm:** Explained in Table 3	$[U_P,U_M]$ in III.

Where:

- C_P and U_P denote being cooperative and being uncooperative, respectively
- For the MNE, C_M and U_M denote being cooperative and being uncooperative, respectively
- Each player's action such as "Partner being cooperative and MNE being uncooperative" is denoted by $[C_P, U_M]$
- The costs of being a cooperative MNE and a cooperative Partner are denoted by A_P
- Benefits due to future market potentials for an MNE, denoted by F_M
- Learning of advanced knowledge/technology from the MNE, denoted by L_P
- The costs due to the situation where only one player is uncooperative, denoted by T_P

identify some important criteria spanning across strategic traits, organizational traits and relational traits. However, although using traits as criteria may help to rank partner candidates, it will have difficulties in judging whether a partner is good enough when some of the criteria are not met. Therefore, it is essential to examine the relative importance of partner selection criteria in terms of the types of fits.

Theory development on the relative importance of partner fits

The partner selection considerations described above can be transformed to a set of equilibrium conditions contingent on different combinations of contexts. Table 7.1 summarizes the six sets of equilibrium solutions. The project complexity is measured by the degree of integrated collaboration efforts required for project success. "Complex" projects are defined by the situation that project success needs both parties to act cooperatively. Project complexity is measured by the opportunity costs of any player's being uncooperative.

Partner selection in scenarios: Local market expansion: complex

Table 7.1 shows that both scenarios have an identical equilibrium path, $[C_P, C_M]$, with $F_M - A_M$ as MNE's payoffs. Implications on partner selection for MNEs, as shown in Table 7.2, can be derived based on their corresponding equilibrium conditions. Note that the learning motivation of the local Partner will be an important contextual factor that affects the equilibrium and subsequent strategy implications. The strategy implications will be discussed as follows.

1. Strategy implications in Local market expansion: Complex: Low learning motivation

- *Focus on strategic fit particularly for higher F_M*: Since the MNE's payoffs in the equilibrium path are $F_M - A_M$, it is essential to select a Partner who owns complementary resources for higher future market potentials, F_M, such as knowledge of local market and social capital, etc.

- *Focus on relational fit as a collaboration goal so as to increase the relational rents of both parties:* Dyer and Singh (1998) argue that relational competitive advantages can be better generated when, among others, there are better knowledge-sharing routines and self-enforced governance mechanism. Here since both partners will be "structurally" being cooperative, better knowledge-sharing routines and effective governance can be achieved more easily. Thus, relational fit should become an objective of collaboration to obtain the higher relational rents, especially when the relational capital may largely increase F_M.
- *Focus on organizational fit for lower A_p:* Lower A_p for a Partner is essential, as the Partner has to act cooperatively but with limited benefits from learning. However, since the MNE's cost of being cooperative, A_M, can be covered by the benefits from future market potentials, it is not essential for the MNE to have a low A_M.

II. Strategy implications in: Local market expansion: Complex: High learning motivation

- *Focus on strategic fit particularly for higher F_M.*
- *Focus on relational fit as a collaboration goal so as to increase the relational rents of both parties.*

Note that the organizational fit for a Partner is not emphasized because the Partner's high learning motivation implies that the benefits from learning can afford higher cost of being cooperative.

Partner selection in scenario: Project profit orientation: Complex

I. Strategy implications given: Project profit orientation: Complex: Low learning motivation

- *Focus on general strategic fit.*
- *Focus on relational fit as a collaboration goal so as to increase the relational rents of both parties.*
- *Focus on mutual organizational fit for lower A_M and A_p:* The focus on a mutual organizational fit is because the MNE has no benefits from future market potentials to cover the higher cost of being cooperative and the Partner has limited benefits from learning.

II. Strategy implications given: Project profit orientation: Complex: High learning motivation

- *Focus on general strategic fit.*
- *Focus on relational fit as a collaboration goal so as to increase the relational rents of both parties.*
- *Focus on mutual organizational fit for lower A_M:* The organizational fit for a Partner is not emphasized because the Partner's high learning motivation implies that the benefits from learning can afford higher cost of being cooperative.

Partner selection in scenario: Local market expansion: Regular and Local market expansion: Simple

Note that the two scenarios here have identical equilibrium solutions and, thus, have the same strategy implications discussed as follows. Again, learning motivation will be a new contextual variable here to influence the equilibrium and strategies.

Table 7.2 Criteria and strategy for partner selection in global projects

	Complex	*Regular*	*Simple*
Local Market Expansion	• Focus on strategic fit particularly for higher F_M • Focus on relational fit as a collaboration goal to increase the relational rents of both parties **&** **If learning motivation is low:** • Focus on organizational fit for lower A_p **If learning motivation is high:** • No additional emphases	• Focus on strategic fit particularly for higher F_M **&** **If learning motivation is low $(L_p<A_p-T_p)$:** • No additional emphases **If learning motivation is high:** • Focus on relational fit as a collaboration goal to increase the relational rents of both parties	• Same emphases as those for the local market expansion with regular projects (on the left)
Project Profit Orientation	• Focus on general strategic fit • Focus on relational fit as a collaboration goal to increase the relational rents of both parties **&** **If learning motivation is low:** • Focus on mutual organizational fit for lower A_M and lower A_p **If learning motivation is high:** • Focus on organizational fit for lower A_M	• Focus on general strategic fit **&** **If the MNE dominates:** • Focus on organizational fit for Partner for lower A_p **If the local Partner dominates:** • Focus on organizational fit for MNE for lower A_M **If no dominant firm:** • If there exists relational fit, both will act cooperatively for better payoffs. Focus on relational fit as a precondition. • If there is no relational fit, both will play mixed strategy and take turn to be cooperative. No additional emphases.	• Focus on general strategic fit

1. Partner selection implications given: Local market expansion: Regular or Simple: Low learning motivation

• *Focus on strategic fit particularly for higher F_M.*

Also note that since the Partner will be acting uncooperatively in this situation, organizational fit for a Partner becomes not essential.

II. Partner selection implications given: Local market expansion: Regular or Simple: High learning motivation

- *Focus on strategic fit particularly for higher F_M.*
- *Focus on the relational fit as a collaboration goal so as to increase the relational rents*: Because the equilibrium resumes back to $[C_P, C_M]$, relational fit as a collaboration objective becomes essential.

Partner selection in scenario: Project profit orientation: Regular

This scenario is the most often observed scenario in global projects. However, this scenario has the most complicated interaction between the players. In this particular scenario, the order of moves does affect the equilibrium solutions. Here, the MNE's move is the opposite of the Partner's move. The insight of this solution is that, because a "regular" project needs only one player being cooperative, the one who moves first will take the first mover advantage acting uncooperatively, forcing the other player to be cooperative to prevent further losses. The question is: Who will be the first mover? Here, the economic rationale is adopted to model the one with dominant power as the first mover. If the local Partner dominates, the local Partner will be modeled as the first mover and the equilibrium path would be $[U_P, C_M]$. Similarly, if the MNE dominates, the equilibrium path would be $[U_M, C_P]$. As such, the power of domination will be added to our model as the fourth contextual variable. The partner selection implications are obtained as shown in Table 7.2.

However, if none of the players dominates, there will be no unique equilibrium path. Normally, this will be a mixed strategy equilibrium, which yields mediocre payoffs. An alternative modeling or strategy for this situation is to assume a game of "promise or punish" for better payoffs, where both players have an agreement ex ante to act cooperatively and if any of the players defects ex post in an event, the other player will play being uncooperative, as a punishment to the one who defects for at least several following events. In other words, each player commits to giving up the individual's best payoffs by behaving opportunistically, $-T_P$ or $-T_M$, in return for obtaining optimal *overall* payoffs, (L_P-A_P, F_M-A_M), and reducing the probability of receiving the worst payoffs, $(-N_P, -N_M)$. Relational capital such as goodwill, trust, past experience, or reciprocal commitments is critical to the success of the "promise or punish" game, or the stability of the equilibrium path aimed ex ante. In this regard, this model may help to explain why the performance effects of relational capital/fit are particularly emphasized in alliance literature.

I. Strategy implications in: Project profit orientation: Regular: MNE dominates

- *Focus on general strategic fit*: If the MNE dominates, the MNE needs not to be cooperative and thus there are no needs for the partner's organizational fit for the MNE. Since this is a regular project, the local Partner's being cooperative is satisfactory to both players.

II. Partner selection implications given: Project profit orientation: Regular: Partner dominates

- *Focus on general strategic fit.*

- *Focus on organizational fit for MNE for lower A_M.*

III. *Partner selection implications given: Project profit orientation: Regular: No dominant firm*

- *Focus on general strategic fit.*
- *If there exists relational fit between the partners, both will act cooperatively for better payoffs. Therefore, the focus will be on relational fit as a precondition.*
- *If there is no relational fit, both will play mixed strategy and take turn to act cooperatively. As such, no additional emphases other than general strategic fit.*

Partner selection in scenario: Project profit orientation: *Simple*

As shown in Table 7.1, the equilibrium path in this scenario is $[U_p, U_M]$ with payoffs ($-N_p$, $-N_M$). Because both players' costs of being uncooperative for a simple project are low, no one has incentives being cooperative given that F_M is small. As we often observed, there will be limited concerns for integrated cooperation when a project is technically simple or the joint venture is simply based on certain uncomplicated resource complementarity such as financial capitals.

Strategy implication:

- *Focus on general strategic fit.*

Discussion

In this section, we present a game theory-based model and discuss the issues in partner selection under multiple contexts. These contextual variables include the MNEs' market strategy, the project complexity, the local Partner's learning motivation, and the power of domination. It is worth noting that, since this model is derived from the behavioral equilibrium of MNEs and local Partners, the model can also provide strategy implications for both MNEs and local Partners. In fact, if the decision of a player cannot satisfy the solution conditions required for the player's counter-party, the collaboration will not be stable and the expected outcome may not be obtained. Another extension of the model is to analyze the interactions between two foreign partners/MNEs who work together in a JV to perform a project. For example, the interactions between MNEs and the firms in the pre-international stage of internationalization process, adopting a "partner following" location strategy, can be analyzed in a similar fashion used in this model.

By analyzing all the solutions and strategy implications in Table 7.2, we find that it seems very difficult for the local Partners to achieve their learning objective in the often-seen scenario where MNEs pursue only project profits in regular projects. In fact, this result is consistent with the observed frequent failures in the local Partners' learning. However, the model can provide viable solutions to this problem. For example, the local Partners may better achieve their learning objective by changing the Project Profit Orientation: Regular scenario to Complex scenario by deliberately altering the project process or context or to Local Market Expansion scenario by creating the "shadows of the future" for MNEs. We also find that strategic fit in terms of complementarity generally is not enough, except for the very simple projects with project profit oriented MNEs.

Conclusions

The global economy has changed drastically how firms operate and compete. Firms have to reconsider with whom they are competing, what their competitive and comparative advantages are, and how they can strengthen their competitiveness to grow. In this chapter, we adopt some non-traditional views and examine three important issues concerning the organization strategies for competing in the move of construction globalization.

The first issue is the internationalization process of construction firms. We present a framework for the internationalization process of construction firms, namely, D-OLI model, and derive implications for internationalization strategies, including the capability development, market locations, and entry mode strategies. According to the model, the development of ownership advantages for global competition will drive the process of internationalization, which includes three stages: the pre-international stage, the pre-MNE stage, and the MNE stage. Location and entry mode strategies appropriate for each stage are also derived in the model. Second, we introduce a conceptual framework for governance structure choices for CJVs. The two distinctive governance structures for CJVs, focusing on the control and coordination aspect of governance, are Separately Managed JVs (SMJ) and Jointly Managed JVs (JMJ). It is hypothesized that the choice of CJV governance structure is largely influenced by four major variables, namely, corporate cultural difference, trust, needs for procurement autonomy, and motivation for learning. The third issue concerns how to select partners for global projects with respect to different contingency factors. We introduced a game theory-based model that analyzes the partner selection problems and suggests strategies under multiple contingency contexts, namely, the MNEs' market strategy, the project complexity, the local Partner's learning motivation, and the power of domination.

For the past two decades, the research on construction internationalization/globalization has been focusing on issues such as the globalization impacts and trends, the key successful factors and risks in internationalization, the emerging international markets or contractors, and the use of joint ventures. For the next two decades, the new challenges of construction globalization will be centered on how to manage MNEs and global projects in a global economy through better strategies that integrate non-traditional concerns such as culture, institutions, organizations, social mechanisms, and sustainability, etc. The three topics introduced in this chapter reflect several of the new challenges. First, the modern construction internationalization process must incorporate the distinct nature of the diverse and fast changing global economy and the characteristics of construction projects and industry. Second, the social mechanism/capital perspective and the RBV are emphasized in studying the contingency contexts of better governance structure for CJVs. Third, in the partner selection model, the fit is defined not only by the complementarity, but also partners' organizational compatibility and relational capital. To summarize, in the modern global economy, more attention from practitioners and more research by scholars will center on the new, non-traditional challenges and issues in order to provide firms more effective strategies and more integrated solutions.

References

Anderson, O. (1993), "On the Internationalization Process of Firms: A Critical Analysis," *Journal of International Business Studies*, 24(2): 209–231.

Badger, W.W. and D.E. Mulligan (1995), "Rationale and Benefits Associated With International Alliances," *Journal Construction Engineering and Management*, ASCE, 121(1):100–111.

Barney, J.B. and M.H. Hansen (1994), "Trustworthiness as a Source of Competitive Advantage," *Strategic Management Journal*, 15(winter): 175–190.

Bon, R. and D. Crosthwaite (2001), "The Future of International Construction: Some Results of 1992-1999 Surveys," *Building Research and Information*, 29(3): 242–247.

Bucklin, L.P. and S. Sengupta (1993), "Organizing Successful Co-marketing Alliances," *Journal of Marketing*, 57(2): 32–46.

Chen, C. (2005), *Entry Strategies for International Construction Markets*, doctoral dissertation, The Pennsylvania State University, PA.

Daft, R.L. (2001), *Organization Theory and Design*, seventh edition, South-Western College, Cincinnati, OH.

Das, T.K. and B.S. Teng (2000), "A Resource-based Theory of Strategic Alliances," *Journal of Management*, 26(1): 31–61.

Dunning, J.H. (1988), "The Eclectic Paradigm of International Production: A Restatement and Some Possible Extensions," *Journal of International Business Studies*, 19(1): 1–31.

Dyer, J.H. and W. Chu (2003), "The Role of Trustworthiness in Reducing Transaction Costs and Improving Performance: Empirical Evidence From the United States, Japan, and Korea," *Organization Science*, 14(1): 57–68.

—— and H. Singh (1998), "The Relational View: Cooperative Strategy and Sources of Interorganizational Competitive Advantage," *The Academy of Management Review*, 23(4): 660–679.

Eisenhardt, K.M. and C.B. Schoonhoven (1996), "Resource-based View of Strategic Alliance Formation: Strategic and Social Effects in Entrepreneurial Firms," *Organization Science*, 7(2):136–150.

Gale, A. and J. Luo (2004), "Factors Affecting Construction Joint Ventures in China," *International Journal of Project Management*, 22(1): 33–42.

Geringer, J.M. (1991), "Strategic Determinants of Partner Selection Criteria in International Joint Ventures," *Journal of International Business Studies*, 22(1): 41–62.

—— and L. Hebert (1989), "Control and Performance of International Joint Ventures," *Journal of International Business Studies*, 20(2): 235–254.

Harrigan, K.R. (1985), "Strategies for Intrafirm Transfers and Outside Sourcing," *Academy of Management Journal*, 28(4): 914–925.

—— (1988), "Joint Ventures and Competitive Strategy," *Strategic Management Journal*, 9(2): 141–158.

Hitt, M.A., M.T. Dacin, E. Levitas, J. Arregle, and A. Borza (2000), "Partner Selection in Emerging and Developed Market Contexts: Resource-based and Organizational Learning Perspectives," *Academy of Management Journal*, 43(3): 449–467.

Ho, S.P. (2009), "A Dynamic OLI Model for the Internationalization Process of Construction Firms," *Research Project Report* #98940, CECI Inc., Taipei, Taiwan.

——, H. Wu, and E. Lin (2009a), "Model for Partner Selection in Global Projects: A Game Theory Analysis," *Proceedings of the LEAD 2009 Specialty Conference: Global Governance in Project Organizations*, South Lake Tahoe, CA.

——, Y. Lin, W. Chu, and H. Wu (2009b), "Model for Organizational Governance Structure Choices in Construction Joint Ventures," *Journal of Construction Engineering and Management*, ASCE, 135(6): 518–530.

——, Y. Lin, H. Wu, and W. Chu (2009c), "Empirical Study of a Model for Organizational Governance Structure Choices in Construction Joint Ventures," *Construction Management and Economics*, 27(3): 315–324.

Horii, T., Y. Jin, and R.E. Levitt(2004), "Modeling and Analyzing Cultural Influences on Project Team Performance," *Journal of Computational and Mathematical Organization Theory*, 10(4): 305–321.

Inkpen, A.C. and M.M. Crossan (1995), "Believing is Seeing – Joint Ventures and Organization Learning," *Journal of Management Studies*, 32(5): 595–618.

Johanson, J. and J. Vahlne (1977), "The Internationalization Process of the Firm-A Model of Knowledge Development and Increasing Foreign Market Commitment," *Journal of International Business Studies*, 8(1): 23–32.

—— and F. Wiedersheim-Paul (1975), "The Internationalization of the Firm-Four Swedish Cases," *Journal of Management Studies*, 12(3): 305–322.

Khanna, T., R. Gulati, and N. Nohria (1998), "The Dynamics of Learning Alliances: Competition, Cooperation, and Relative Scope," *Strategic Management Journal*, 19(3),: 193–210.

Levitt, R. (2007), "CEM Research for the Next 50 Years: Maximizing Economic, Environmental, and Societal Value of the Built Environment," *Journal of Construction Engineering and Management*, ASCE, 133(9): 619–628.

Luo, J. (2001), "Assessing management and performance of Sino-foreign construction joint ventures," *Construction Management and Economics*, 19(1): 109–117.

Luo, Y. (1997), "Partner Selection and Venturing Success: The Case of Joint Ventures with Firms in the People's Republic of China," *Organization Science*, 8(6): 648–662.

Mahalingam, A. and R.E. Levitt (2007), "Institutional Theory as a Framework for Analyzing Conflicts on Global Projects," *Journal of Construction Engineering and Management*, ASCE, 133(7): 517–528.

Mayer, R.C., J.H. Davis, and F.D. Schoorman (1995), "An integrative model of organizational trust," *Academy of Management Review*, 20(3): 709-734.

Meschi, P.X. (1997), "Longevity and Cultural Differences of International Joint Ventures: Toward Time-based Cultural Management," *Human Relations*, 50(2): 211–228.

Mjoen, H. and S. Tallman (1997), "Control and Performance in International Joint Ventures," *Organization Science*, 8(3): 257–274.

Mohamed, S. (2003), "Performance in International Construction Joint Ventures: Modeling Perspective," *Journal of Construction Engineering and Management*, ASCE, 129(6): 619–626.

Morris, J.M. (1998), *Joint Ventures: Business Strategies for Accountants*, second edition, J. Wiley & Sons, New York.

Naylor, J. and M. Lewis (1997), "Internal alliances: Using joint ventures in a diversified company," *Long Range Planning*, 30(5): 678–688.

Ngowi, A.B. (2007), "The Role of Trustworthiness in the Formation and Governance of Construction Alliances," *Building and Environment*, 42(4): 1828–1835.

——, E. Pienaar, A. Talukhaba, and J. Mbachu (2005), "The Globalisation of the Construction Industry – a Review," *Building and Environment*, 40(1): 135–141.

Ofori, G. (2003), "Frameworks for Analysing International Construction," *Construction Management and Economics*, 21(4): 379–391.

Ozorhon, B., I. Dikmen, and M.T. Birgonul (2007), "Using Analytic Network Process to Predict the Performance of International Construction Joint Ventures," *Journal of Management in Engineering*, 23(3): 156–163.

Ozorhon, B., D. Arditi, I. Dikmen, and M.T. Birgonul (2008), "Effect of Partner Fit in International Construction Joint Ventures," *Journal of Management in Engineering*, 24(1): 12–20.

Park, S.H. and G.R. Ungson (1997), "The Effect of National Culture, Organizational Complementarity, and Economic Motivation on Joint Venture Dissolution," *Academy of Management Journal*, 40(2): 279–307.

Powell, W.W. (1990), "Neither Market nor Hierarchy: Network Forms of Organization," *Research in Organizational Behavior*, 12: 295–336.

Sarkar, M.B., R. Echambadi, S.T. Cavusgil and P.S. Aulakh (2001), "The Influence of Complementarity, Compatibility, and Relational Capital on Alliance Performance," *Journal of the Academy of Marketing Science*, 29(4): 358–373.

Saxton, T. (1997), "The Effects of Partner and Relationship Characteristics on Alliance Outcomes," *The Academy of Management Journal*, 40(2): 443–461.

Shah, R.H. and V. Swaminathan (2008), "Factors Influencing Partner Selection in Strategic Alliances: The Moderating Role of Alliance Context," *Strategic Management Journal*, 29(5): 471–494.

Sillars, D.N. and R. Kangari (1997), "Japanese Construction Alliances," *Journal of Construction Engineering and Management*, ASCE, 123(2): 146–152.

—— and —— (2004), "Predicting Organizational Success Within a Project-based Joint Venture Alliance," *Journal of Construction Engineering and Management*, ASCE, 130(4): 500–508.

Ulrich, D. and D. Lake (1991), "Organizational Capability: Creating Competitive Advantage," *Academy of Management Executive*, 5(1): 77–92.

Warszawski, A. (1996), "Strategic Planning in Construction Companies," *Journal of Construction Engineering and Management*, ASCE, 122(2): 133–140.

Wernerfelt, B. (1984), "A resource-based view of the firm," *Strategic Management Journal*, 5(2): 171–180.

Zaheer, A., B. McEvily, and V. Perrone (1998), "Does Trust Matter? Exploring the Effects of Interorganizational and Interpersonal Trust on Performance," *Organization Science*, 9(2): 141–159.

8 Strategic issues in entering emerging markets

I. Dikmen, M.T. Birgonul, and C. Anac

Introduction:

Although international construction activities help construction firms to mitigate the risk of adverse change in domestic market conditions, taking advantage of opportunities in emerging markets and increasing competencies as a result of lessons learned in international markets poses serious threats due to instability of competitive conditions and difficulties specific to international construction projects. Mahalingam and Levitt (2007) argue that in addition to the complexities present in construction projects, global projects are distinct from domestic construction projects in that global projects involve interactions among individuals, organisations and agencies from diverse national backgrounds and cultural contexts. Global projects carried out in emerging markets are even more complex than others, as the "instability" characterised in these markets when coupled with "differences" between home and host country practices tend to create major problems that significantly affect the success of construction projects. The purpose of this chapter is to discuss how strategic issues should be taken into account when entering emerging markets and propose a generic market entry decision making process for organisations that encompasses these issues.

Market entry decisions

Global expansion is a strategic choice. Market selection that follows a global expansion decision requires an analysis of the competitive environment as well as the firm's resources and capabilities to achieve market fit. Porter's five forces model (1980) relies on the analysis of a company's external competitive environment as the basis of strategic choice. Porter's five forces, namely, threat of new entrants/competitors, level of rivalry between competitors, threat of substitute products, bargaining power of suppliers and bargaining power of customers determine the level of competition and profitability in a market. Companies should find a strategic position that is less affected from these forces (Figure 8.1a).

The diamond framework proposed by Porter (1990), has enhanced the understanding of sources of national comparative advantage (Figure 8.1b) The diamond framework was proposed to explain why companies hosted in some countries have been successful in penetrating foreign markets in some product areas but not in others, and also why some countries have been able to attract the participation of foreign-owned firms in some value added activities but not in others. Regarding demand conditions, Porter believes that home demand has a considerable influence on competitive advantage, and he presents the composition, the size and pattern of growth, and the internationalisation of home demand as three broad attributes of it. The existence of internationally competitive related and supporting

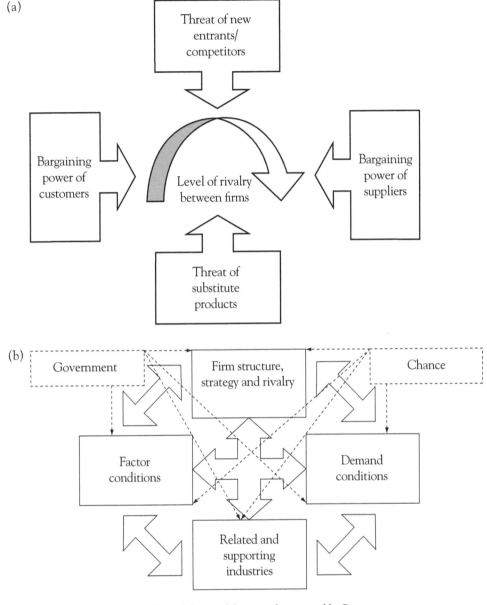

Figure 8.1 Five forces model and diamond framework proposed by Porter.

industries in a nation, according to Porter, is also an important determinant of creation and sustainability of competitive advantage. However, Grant (1991) argues that diamond framework lacks clarity on hierarchy of resources for competitive advantages in terms of sustainability.

Another prevailing theory in strategic management literature that is adapted to international business is the resource-based view (RBV). The resource-based view is grounded in the perspective that a firm's internal environment, in terms of its resources and capabilities, is more critical to the determination of strategic action than the external environment – that a firm's competitive strategy is critically influenced by its accumulated resources. The RBV suggests that the resources possessed by a firm are the primary determinants of its performance, and these may contribute to a sustainable competitive advantage of the firm (Hofer and Schendel 1978). According to Barney *et al.* (2001), the concept of resources includes all assets, capabilities, organisational processes, firm attributes, information and knowledge controlled by a firm that enable the firm to conceive of and implement strategies that improve its efficiency and effectiveness. The RBV treats enterprises as potential creators of value-added capabilities. It focuses on the idea of costly-to-copy attributes of the firm as sources of business returns and the means to achieve superior performance and competitive advantage. According to the RBV, global expansion is a strategic choice that has an objective of the enlargement of the resource base and the exploitation of the existing ones (Yeniyurt *et al.* 2005). A major contribution of the RBV is that it provides valuable suggestions for a firm to focus on those firm-specific internal resources (Flanagan *et al.* 2007). However the basic pitfall for RBV theory is the assumption of a fair and homogenous competitive environment. Especially in emerging or developing economies where the transition of legislation and legal structures cause a volatile environment, it results in a comparatively unfair competitive environment. Porter's work and RBV are considered as complementing views despite their remote bases. Both the rules of competition and company strengths and weaknesses shall be taken into account while taking market entry decisions.

Gunhan and Arditi (2005) provide a third perspective that hypothesises that international expansion decisions shall be governed by conducting internal readiness tests, external readiness tests and country specific analyses. The first test covers the company strengths and resource analysis to assess whether the construction company is strong enough for international expansion. The second test focuses on the threats that shall be faced by the construction company while operating in foreign countries. The third test involves the target country related issues during which the potential difficulties and differences are identified and classified (such as country-specific economical and political issues). Finally, market entry mode is selected considering the results of the mentioned analyses. It is apparent that market entry decision making is an iterative process of conducting company and market level analyses to generate viable strategies to achieve market fit. Figure 8.2 demonstrates a generic process for entering international markets.

As depicted in Figure 8.2, the internationalisation decision is the starting point of the market entry decision making process. Horst (1972) hypothesises that the firm runs through its opportunities in the domestic market before incurring the transaction cost of going abroad. The firm takes on ventures abroad only after it has accumulated some critical mass of assets. Ngowi *et al.* (2005) confirms that a company must have acquired some important capabilities to enter the international market. There are several studies for defining strategic motives for internationalisation decisions (Chittor 2007; Dahringer 1991; Dunning and Narula 2004; Helpman *et al.* 2004; Nachum *et al.* 2001). These studies mainly focus on manufacturing firms; however three main motives can be applied to construction (Flanagan *et al.* 2009):

- **Market seeking**: It is a primary (and defensive) motive for expanding business overseas. The rise of market seeking strategies by big companies goes back to the post-World War II era. However, with globalisation in the 1980s, the need for greater international business became vital for the survival of a company. Market seeking is also influenced by the desire to distribute the workload risk.

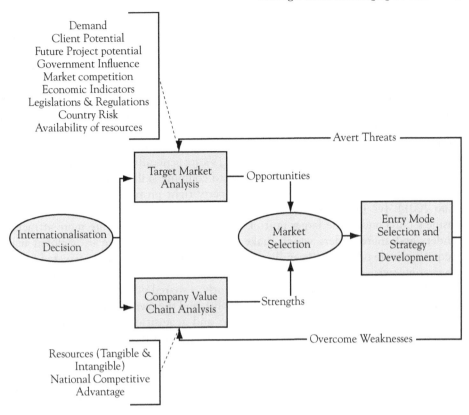

Figure 8.2 A generic process model for entering an international market.

- **Resource seeking**: With the breakdown of national borders and an increase in international work, the intense competition is in cutting the cost of resources. A recent example of this is the resource transfer between Europe and Asia.
- **Efficiency seeking**: It is mainly adopted by European companies. Process outsourcing is a common way of business process improvements. Companies seeking to achieve asset diversification and risk minimisation, move into foreign markets. Efficiency seeking requires an advanced organisational structure therefore mergers, acquisitions or partnership formations are common ways to strengthen core competencies.

Construction companies choose overseas markets where they have competitive advantage, based on firm and national advantages like other business enterprises (Seymour 1987). Firm-specific advantages include: the firm's name, which embodies its reputation, experience and expertise; and firm size, which relates to its resources (Ofori 2003). National advantages include: national currency; geographical proximity to market(s); historical, political, language, cultural and economic relationships between the home and host countries; foreign direct investment by home-country enterprises; and strengths of intersectoral linkages within the home country's economy. The process of company value chain analysis should be comprised of identification of all firm and national advantages and their impacts on the value chain. Target market analysis requires identification of market factors such as risk, demand, competitive forces, government and availability of resources. Opportunities in the market and availability of necessary competencies drive the market selection decision whereas strategies are generated in an iterative process to minimise vulnerability to market threats and overcome initial weaknesses. Within this respect, the entry mode selection is a strategic decision. There are various studies that discuss alternative modes of entry into international markets. Ling *et al.* (2008) point out that one of the possible entry modes for architectural, engineering and construction firms is setting up a subsidiary (representative office). Yet, strategic partnerships and joint ventures (JVs) are the most common entry modes to undertake projects in international markets. JVs are generally established between a local and foreign company when there is a need to combine skill or financial resources with local know-how or when host-country property right protection legislations are strict (Luo 2001). Ive (1994) states that large firms can spread or pool the risks of many business units under one larger unit of ownership and that they are more attractive to lenders and shareholders which make them more competitive in a global market. Acquisitions are takeovers (friendly or hostile) of one company over a targeted company, whereas mergers are more complex and aim to form a bigger company by combining skill bases, financial resources and, more importantly, a shared vision and objectives. Mergers are harder and more time consuming to establish and the ownership structure of the firms is crucial. The basic motive behind mergers and acquisitions is financial performance improvement. Mergers and acquisitions among large construction companies have been popular in the last 10–20 years, mainly in advanced industrialised countries like the United States and European countries. Common entry modes are summarised in Table 8.1. Less developed countries enforce high entry barriers. Generally, JVs are popular in less developed and developing countries because of the benefits of technology transfers, risk sharing, job creation and capital inflows (Ling *et al.* 2005).

In this chapter, the strategic issues considered within the generic process explained in Figure 8.2 will be examined by focusing on the distinctive characteristics of emerging markets. The threats specific to emerging markets as well as required company competencies and strategies to achieve market fit will be discussed in the next section.

Distinctive characteristics of emerging markets

The term, *emerging market* was first introduced in early 1980s by Antoine W. Van Agtmael of the International Finance Corporation of the World Bank. The concept of emerging markets signified a business phenomenon which is not fully described or bounded by regional and economical strength. The United Nations (2005) states the top six most attractive global business locations where five are emerging economies; namely, China, India, Russia, Brazil and Mexico. These countries represent the role of economic powerhouses with their significantly large population, resource base and growing/potential large market. Emerging countries are those restructuring their economies along market oriented lines and aiming to offer better opportunities in trade, technology transfer and foreign direct investment. Heakal (2003) characterises emerging market economies (EME) as *transitional* due to their process from a closed economy to an open market economy. Emerging markets are sometimes defined as those countries that are developing and resource-rich (Han *et al.* 2010). The distinctive characteristics of emerging markets are identified as economic and social dynamism, high level of competition, pervasive market failure and institutional differences (Enderwick 2009).

Table 8.1 Market entry modes

Modes of entry	Advantage
Subsidiary/representative office	No risk sharing, ease of establishment, dependent on regulations in the host country, low local fit
Joint ventures	Easy to establish, great importance on partner selection criteria
Acquisitions of a local competitor	Requires financial resources, dependent on host country regulations, high risk
Mergers	Requires financial resources, negotiations may take very long time, quick access to local knowledge, long term commitment

Literature findings on the distinctive characteristics of emerging markets from developing and developed markets that should be considered during market entry decisions by construction professionals are summarised in Table 8.2. As it is clear from Table 8.2, the emerging market characteristics lie in between those of developing and developed markets, which pinpoints the existence of "instable" and "dynamic" conditions that result in high level of "uncertainty" and "vagueness".

Some strategic issues, specific to emerging markets, as depicted in Table 8.2 are summarised below.

Motives for entering emerging markets

Motives are mainly high demand (attractive market opportunities), low cost sources and learning opportunity. Enderwick (2009) discusses that lessons learned in emerging markets have a competitive value in other developing economies as well as developed country markets. The major learning opportunity provided by emerging markets is about the management of change which may be particularly applied in other emerging markets that might be expected to have a similar change. The lessons learned in emerging markets affect the business strategy of international firms creating a major source of competitive advantage during global competition.

Market conditions

The specific market conditions that should be taken into account during market analysis are:

Potential return

Emerging economies are attractive for business because of their sometimes large and often fast-growing markets, and because they provide access to resources, notably raw materials and labour not available at the same cost as in mature economies (Estrin and Meyer 2004). The definition of emerging markets requires that national economic growth indicators be high. Yet the transition from interdependency to global interconnectivity results in critical instabilities. The nature of these markets engage higher risks on projects, however, due to domestic dependencies on local low cost-high quality resources, may result in higher return rates.

Table 8.2 Distinctive characteristics of emerging markets

	Developing markets	*Emerging markets*	*Developed markets*	*References*
Country Conditions				
Political conditions	Unstable and highly influential	Unstable and highly influential	Stable with low influence	Kim *et al.* 2009; Hitt *et al.* 2000; Yng *et al.* 2005; Mody 2004; Füss 2001; Diamonte *et al.* 1996; Salomons and Grootveld 2003; Harvey 1995; Glen and Singh 2004; Luo 2001; Nakata and Sivakumar 1997
Economic conditions	Unstable and poor indicators	High economic growth with unstable indicators	Low economic growth with stable indicators	Kim *et al.* 2009; Hitt *et al.* 2000; Ling *et al.* 2005; Mody 2004; Nakata and Sivakumar 1997
Legislations and regulations	Highly immature	Immature and unstable legal system with low enforceability	Mature and stable legal system with high enforceability	Glen and Singh 2004; Luo 2001; Hitt *et al.* 2000; Nakata and Sivakumar 1997
Sociocultural risks	High	Moderate-high	Low	Kim *et al.* 2009; Nakata and Sivakumar 1997
Knowledge/ property protection systems	Weak	Weak	Strong	Han and Diekmann 2001; Luo 2001; Hitt *et al.* 2000; Nakata and Sivakumar 1997
Government intervention	High	High	Fair	Füss 2001; Luo 2001; Han and Diekmann 2001
Market Conditions				
Market demand	Low demand with lack of initiatives	High demand with rapid growth	Low demand with stable growth (mature)	Hitt *et al.* 2000; Füss 2001; Nakata and Sivakumar 1997
Rate of return	Unpredictable	High rate with low predictability	Relatively low rate with high predictability	Moreno and Olmeda 2007; Salomons and Grootveld 2003; Harvey 1995; Mody 2004; Füss 2001
Local knowledge and relations	Highly critical	Highly critical	Less critical	Yng *et al.* 2005; Moreno and Olmeda 2007; Harvey 1995; Hitt *et al.* 2000; Luo 2001

Table 8.2 (continued)

	Developing markets	Emerging markets	Developed markets	References
Interconnectivity with global markets and regulations	Low	Low-moderate	High	Salomons and Grootveld 2003; Füss 2001; Harvey 1995; Korajczyk 1996; Glen and Singh 2004; Hitt *et al.* 2000
Competition	Price-based	In transition between price- and technology-based	Technology-based	Ling *et al.* 2005; Glen and Singh 2004
Competency of local firms	Low	Low (but improving)	High	Ling *et al.* 2005

(row label, rotated: Market Conditions)

Construction demand

Emerging markets provide higher income growth than developed nations do. This fact results in increasing purchasing power among consumers and consecutively higher demand. Demand is very high especially on infrastructure and development projects. Infrastructure projects are generally long-lasting construction projects which grant a continuous workload for construction firms.

Competition

The global market competition rules are not fully applied. The competition in these markets is mostly driven by low cost and traditional processes have not fully disappeared. However, competition is shifting from conventional price-based competition to a more complex structure where non-price factors become more critical. Even in the lowest bid opportunities, more clients are considering the contractors' ability to offer additional services such as technology transfer and project development, financing and management expertise (Han *et al.* 2010).

Risk

Risks in the emerging markets are high due to instability of political and economic conditions as well as the immaturity of legal systems. When compared with developed countries, construction projects carried out in emerging markets involve a higher degree of risk due to their vulnerability to instable country conditions (mainly economic and political). However, Cavusgil *et al.* (2002) signify the paradigm shift of international business in emerging countries where the risks are becoming increasingly manageable. Enderwick (2009) mentions that risk in emerging countries does not follow the patterns apparent in the least developed economies where political and economic risks are widespread. Rather, they stem from the transition process and economic, social and policy changes.

Competency of local firms

Governments in emerging countries usually make it mandatory for foreign firms to enter into joint venture deals with local partners (Han *et al.* 2010). The local companies in emerging markets are also in a transition state. Although their managerial capabilities and technological know-how are not advanced as in firms from developed countries, increasing inward and outward international construction activities lead to rapid growth of local firms. Joint venturing with multinationals in large construction projects results in technology transfer as well as increases in experience of managing large scale projects. Thus, the competencies, resources and capabilities of emerging market enterprises are evolving.

Potential clients

Potential clients are private organisations or mainly public institutions which are not fully institutionalised, which creates a major risk for foreign contractors due to the vagueness of the client's work practices and regulations.

Required company competencies

The competencies necessary to achieve competitiveness in emerging markets are mainly managing change, managing risk and understanding of institutional differences. The problems faced by foreign companies doing business in emerging markets originate from the lack of familiarity with the local environment, which is denoted as "liability of foreignness" by Javernick-Will and Levitt (2010). Local knowledge is necessary to mitigate the liability of foreignness identified by Javernick-Will and Levitt for the contractors are knowledgeable about operating laws, material and labour quality, availability and cost, logistics, work practices and approval processes. There is clearly a need for local knowledge and local partner contribution. Interpersonal or interorganisational relationships and personal connections are more important than legal standards in emerging markets (Hitt *et al.* 2000). Mahalingam and Levitt (2007) argue that the problem of cross-national conflicts on global projects is a result of regulative, normative and value-based institutional differences. Consequently, joint venturing and partnership with local companies becomes a vital strategy for gathering local knowledge, maintaining good relations with the local bodies and minimising institutional differences where one of the critical success factors for international construction turns out to be choosing the right local partner and good partner relations. However, joint venturing brings its own risks due to cultural differences between the companies as well as different working practices.

In the next section, lessons learned from an international construction project carried out in Turkey will be discussed to demonstrate the impact of emerging market risks on the success of a construction project when coupled with lack of experience of a foreign contractor that recently entered the Turkish market.

Case study

Turkey as an emerging market

Construction is one of the leading sectors in Turkey that contributes to the national economy and employment due to its linkages with over 200 supplementary sectors. Construction contributes significantly to employment and gross national product (GNP). Construction investment corresponds to 50 per cent of total investment and has a direct share of 5 per cent and an indirect share of 33 per cent in GNP distribution (Birgonul *et al.* 2009). The construction industry provides employment to about 1.4 million people, representing almost 6 per cent of the total employment in Turkey. Construction investments cover 60 per cent of total fixed investments in Turkey. The share of public construction investments in this proportion is 30 per cent.

There are over 10,000 contractors in the Turkish construction sector; however there are no criteria for contractor certification. This contractor base mainly consists of small- and medium–sized enterprises where the top 150 domestic contractors cover 70 per cent of the national construction volume. The economic crises of 2000 and 2001 have been driving forces for Turkish contractors to attack foreign markets, which increased the number of companies who have become aware of the requirements for global competition (Elci 2003). Currently, there are 23 Turkish contractors among the leading 225 international construction companies in the *Engineering News-Record* list (ENR 2008). After the United States and China, Turkey has the highest number of international contractors in the ENR list. As well as the national comparative advantage stemming from the existence of supporting industries (mainly, high quality materials), low cost labour, culture, and religious similarities with countries that have high construction demand, competencies have been developed as a result of the lessons learned in the domestic market as well as international markets by joint venturing with foreign companies.

The demand for construction works in Turkey is high and concentrates on infrastructure projects. Within the agenda of the Turkish government, there are urgent energy and transportation projects that are planned to be realised by the involvement of foreign contractors and investors. However, there is a low level of institutionalisation in Turkey with weak financial structures in public and private bodies. Turkish public organisations, in general, are renowned for their tedious bureaucratic procedures, which involve high amounts of paperwork, long waiting periods, and many redundant formalities. It is obvious that setting a less hierarchical system for a more time-efficient process requires changes in the present one; such changes are anticipated to be difficult to implement. Although major constitutional and legislative reforms are being made in Turkey during the accession negotiations between Turkey and the European Union (EU), there is still considerable work to do as Turkey has to adopt and implement the whole body of EU legislation and it is still uncertain whether Turkey will be able to achieve its goal of joining the EU. Thus, Turkey is characterised by its high potential but also uncertain business environment resulting in considerable country risks for foreign investors.

Newcomer in an emerging market

The case study is based on the experience of a European contractor which is one of the leading providers of construction services in Central and Eastern Europe. It employs over 45,000 people at more than 500 locations and attains a building performance of more than ten billion euros. The case study project is their first job in Turkey. The sample project is an energy project that has been executed in the north-east region of Turkey. The project has been financed and delivered according to a private agreement between two governments including another hydro-electrical power plant (HEPP). The project consists of civil works, mechanical and electrical instrumentation works where civil works are executed by an international consortium between Turkish and European companies. The names of the companies and project have been disguised for the sake of confidentiality. Three key personnel who worked at the case study project at the management level were interviewed to discuss potential problems that were experienced in this project due to threats specific to country conditions. The interviews were conducted nearly one year after the project was over and, currently, negotiations between the project participants continue to share the significant cost and time overrun experienced in the project.

The purpose of this case study is not to present a generic list of problems that may arise in construction projects carried out by foreign contractors in emerging markets. Rather, the aim is to discuss some strategic issues that should be considered during the market entry decision-making process. Moreover, within the context of this case study, only the problems that were experienced as a result of emerging market conditions will be highlighted. The basic characteristics of the emerging markets which are "instability of country conditions" resulting in "changes" and "differences between firms from developed countries and local firms" are discussed by giving examples from the case study project.

Motives for market entry

As the European contractor entered the Turkish market as a result of a pre-agreement between the two countries; competency and market analysis was not carried out in detail. One of the interviewees indicated that project opportunities were considered whereas some of the country-related issues were overlooked while entering the Turkish construction market. This resulted in a lack of enough knowledge about market practices and local conditions which further resulted in significant problems.

Market conditions

The instability of country conditions resulted in the following "changes" and the project was affected adversely by these changes due to vagueness in some of the contract clauses. Also, "differences" between the project participants and interpretation of some technical terms created a major source of problems in the project.

Economic instability

Due to the multinational nature of the project, the payments in the project were made in different currencies and there was an escalation formula within the contract to protect the contractor against changes in exchange rates. Ongoing economic volatility in Turkey resulted in an economic crisis in 2001 which caused significant fluctuations in exchange rates. A national downturn in economic indicators affected the overall economy together with supporting sectors of construction. The currency dependency of progress payments resulted in remarkable and unpredictable losses for the European contractor. Contractual contingency values were not sufficient to cover these losses. The European contractor had to compensate the losses due to a change in exchange rates using its own financial sources for the sake of timely completion of the project.

The problem of instability of economic conditions was significant not only due to the high rate of change in exchange rates but also in the vagueness of the escalation formula depicted in the contract. The escalation formula had two parts. The second part of the escalation formula gave a negative value for a specific period of time. It is clearly stated in the contract that the minimum value for the first part should be taken as zero, if it is negative. But a similar condition for the second part was not specified. The client made deductions rather than escalation, which turned into a claim issue between the contractor and the client. Consequently, in construction projects, it is not only the instability of country conditions that affect the project but also the level of vulnerability to those conditions due to lack of enough contractual protection of the contractor against "changes".

Political instability

As a result of political changes, the staff working in the client organisation changed frequently resulting in a problem of sustaining good relations between the client and contractor. The high staff turnover rate caused problems in approval and permit procedures as well as progress payments. Moreover, the European contractor had little knowledge about the bureaucratic issues in Turkey and did not have an experienced local staff to deal with issues. Thus, changes in staff in the client organisation resulted in significant bureaucratic delays when coupled with lack of experience of the European company on these issues.

Change in international relations

A significant dispute was experienced between the Turkish government and the government of the European contractor during negotiations of Turkey to join EU. This resulted in an adverse attitude of the client organisation towards the contractor which caused delays in approval procedures and progress payments. Also, the conservative sociocultural environment of the region where the project was being carried out showed distrustful behaviours towards the contractor.

Moreover, the major "differences" between the regulations and norms prevailing in Turkey and the contractor's host country resulted in significant disputes between the project participants, some of which are discussed below.

Difference in definitions about work practices

The payment method of the project was a combination of lump-sum and unit-prices. For some items, lump-sum prices were fixed whereas for the part of the work having uncertain quantities, unit prices were applied. Change orders by the client related to the portion of the work that would be paid in lump-sum prices, and major problems were experienced within the project as the interpretation of lump-sum contract in the Turkish practice is different from the general practice. If the quantities are less than the reservation amounts, the deductions are made based on the lump-sum prices, if they increase the additional part is paid using the government's unit-prices. The unit-prices were significantly lower than the lump-sum prices, creating a significant loss for the contractor. This issue also became a claim issue between the client and contractor.

Cultural differences

Interpersonal and interorganisational relations and personal contacts with the client play a more important role than standards in emerging markets, and Turkey is not an exception. As far as the time and financial constraints of the project are concerned, the European contractor failed to initiate the required communication. The parties could not communicate to solve problems regarding the technical, financial and contractual issues because of cultural differences, language problems and lack of experienced staff in the company who could develop good relations with the personnel in the client organisation.

Required competencies

The European contractor was one of the most competitive companies within the international construction market due its technical expertise and technological know-how in complex engineering projects. Also, its major source of competitive advantage stemmed from its national comparative advantage as the financing of the project was secured by funds coming from its home country. Thus, at the market entry stage, Turkey was an attractive market for the company due to high market demand and its high competitiveness. However, lack of local knowledge, relations with local client organisations and staff experienced in conducting work in emerging markets, meant the company did not have the required competencies to sustain its competitive advantage in the Turkish construction market. Individuals involved with the project mentioned that the company failed to adapt their home country competencies to the requirements of the Turkish construction sector. Their focus on the execution process which emphasised finishing the project at the highest quality within the required time scale resulted in overlooking some of the contractual issues and not keeping enough documentation throughout the project. The lack of in-place contract management and poor understanding of local legislation requirements resulted in poor claim management towards the end of the project.

In the case study project, partner selection schemes were not fully investigated by the foreign contractor when entering the Turkish market. The local partner was a medium-large scale company which was also very experienced in large scale engineering projects and had significant international expertise as well as domestic market experience. It had very good relations with local clients and government bodies. The local partner identified the risks and mitigated them effectively by referring to certain clauses of local regulations. However, the European contractor failed to realise that they should learn from their local partner about local regulations and how to communicate with the client. The consortium partners did not communicate well and have a common strategy for the project execution which resulted in conflicts, even claims between the partners.

After the unsuccessful practices in this project, the company continued to search for new projects in the Turkish market due to their long term commitment in the region. Turkey was still an attractive market for the company as Turkey has shifted its investment focus from housing to major infrastructure and development projects in the recent years, most of which required a foreign investment partner as a part of European Union (EU) accession processes. However, government involvement limited the property ownership rules in these projects while the European contractor failed to proactively initiate investment paths from existing international connections within their corporate structure. The primary business model implemented for the continued existence in the market was a sole venture subsidiary mode. Competitive strategy of the company was based on differentiated technical competencies over local competitors. However, Turkey as an emerging market imposes competition rules based on lowest price and competition in the market is very high due to the high volume of local competitors mainly competing on low price.

The case study project demonstrates the impact of emerging market threats on construction projects. Lessons learned from this project pinpoint the importance of consideration of market as well as competency related issues during market entry decision making. It is apparent that although the source of problems in emerging markets is due to instability of country conditions and differences between the companies as well as practices in host and home markets, the impact of these conditions on a project may change considerably depending on the strategies used by the foreign companies and their ability to deal with market threats.

Discussions

A decision to enter an emerging market should consider market as well as company conditions and effective strategy formulation to avert threats and overcome weaknesses. However, it is also apparent that although a very detailed market analysis may be conducted before entering a market, there still may be problems due to the distinctive characteristics of emerging markets that are unstable and unpredictable. Consequently, the strategy of firms when entering emerging markets should not focus on prediction of market conditions, rather they should mainly try to reach local knowledge to minimise vagueness, which is usually hard to achieve. A strong local partner that complements the initial weaknesses of the foreign company and minimises the impact of differences; contractual protection and good risk management plan against market instabilities; maintaining good relations with local agencies; and an adaptive management that continuously integrates the lessons learned in the market to current strategies appear to be critical success factors. More research and exploratory case studies are necessary to enlighten the critical success factors for carrying out international projects in emerging markets. Moreover, an analysis of the impact of lessons learned in emerging markets on the competencies of firms and reverse knowledge transfer mechanisms shall be investigated so that opportunities and threats related with emerging markets can be discussed from a wider perspective.

References

Barney, J., Wright, M., and Ketchen Jr, D.J. (2001), "The resource-based view of the firm: Ten years after 1991", *Journal of Management*, 27(6): 625–641.

Birgonul, M.T., Dikmen, I., and Ozorhon, B. (2009), "The impact of reverse knowledge transfer on competitiveness: The case of Turkey", in L. Ruddock (ed.), Chapter 11, *Economics for the Modern Built Environment*, Abingdon: Taylor & Francis, pp. 212–228.

Cavusgil, S.T., Ghauri, P.N. and Agarwal, M.R. (2002), *Doing Business in Emerging Markets: Entry and Negotiation Strategies*, Sage Publications, Thousand Oaks, CA/London.

Chittor, R. (2007), *Services Internationalization: A Review and Reflection*, India Institute of Management, Calcutta.

Dahringer, L.D. (1991), "Marketing services internationally: barriers and management strategies", *The Journal of Services Marketing*, 5(3): 5–17.

Diamonte, R.L., Liew, J.M. and Stevens, R.L. (1996), "Political risks in emerging and developed markets", *Financial Analysts Journal*, 52(3): 71–76.

Dunning, J.H. and Narula, R. (2004), *Multinationals and Industrial Competitiveness: A New Agenda*, Edward Elgar, Cheltenham.

Elci, S. (2003), "Innovation policy in seven candidate countries, the challenges: Innovation Policy Profile: Turkey, Final Report, March 2003, Volume 2.7, Enterprise Directorate - General Contract", online, available at: ftp://ftp.cordis.europa.eu/pub/innovation-policy/studies/turkey_ final_report_ march_2003.pdf [accessed 15 November 2009].

Enderwick, P. (2009), "Large emerging markets (LEMs) and international strategy", *International Marketing Review*, 26(1): 7–16.

ENR (*Engineering News-Record*) (2008), *The Top 225 International Contractors*, online, available at: http://enr.construction.com/people/topLists/topIntlCont/topIntlcont_1-50.asp [accessed 20 November 2009].

Estrin, S. and Meyer, K. (2004), *Investment Strategies in Emerging Markets*, Edward Elgar, Cheltenham.

Flanagan, R., Jewell, A. C. and Anac, C. (2009), *A Review of Growth for Construction Service Companies in Global Markets*, Fifth International Conference on Construction in the 21st Century (CITC-V), Collaboration and Integration in Engineering, Management and Technology, Istanbul, pp. 950–958.

Flanagan, R., Lu, W., Shen, L. and Jewell, C. (2007), "Competitiveness in construction: a critical review of research", *Construction Management and Economics*, 25(9): 989–1000.

Füss, R. (2001), *The Financial Characteristics between Emerging and Developed Equity Markets*, online, available at: www.ecomod.net/Conferences/EcoMod2002/papers/fuss.pdf [accessed 22 November 2009].

Glen, J. and Singh, A. (2004), "Comparing capital structures and rates of return in developed and emerging markets", *Emerging Markets Review*, 5(2): 161–192.

Grant, R.M. (1991), "Porter's 'Competitive Advantage of Nations': an assessment", *Strategic Management Journal*, 12(7): 535–548.

Gunhan, S. and Arditi, D. (2005), "International expansion decision for construction companies", *Journal of Construction Engineering and Management*, 131(8): 928–937.

Han, S.H. and Diekmann, J. E. (2001), "Approaches for making risk-based go/no-go decisions for international projects", *Journal of Construction Engineering and Management*, 127(4): 300–308.

Han, S.H., Kim, D.Y., Jang, H.S. and Choi, S. (2010), "Strategies for contractors to sustain growth in the global construction market", *Habitat International*, 34(1): 1–10.

Harvey, C.R. (1995), "Predictable risk and returns in emerging markets", *The Review of Financial Studies*, 8(3): 773–816.

Heakal, R. (2003), "What is an emerging market economy?" *Investopedia Newsletter*, online, available at: www.investopedia.com/articles/03/073003.asp [accessed 25 November 2009].

Helpman, E., Melitz, M.J. and Yeaple, S.R. (2004), "Export versus FDI with heterogeneous firms", *The American Economic Review*, 94(1): 300–316.

Hitt, M.A., Dacin, M.T., Levitas, E., Arregle, J.-L. and Borza, A. (2000), "Partner selection in emerging and developed market contexts: resource-based and organizational learning perspectives", *The Academy of Management Journal*, 43(3): 449–467.

Hofer, C.W. and Schendel, D. (1978), *Strategy Formulation: Analytical Concepts*, West, St. Paul, MN.

Horst, T. (1972), "Firm and industry determinants of the decision to invest abroad: an empirical study", *The Review of Economics and Statistics*, 54(3): 258–266.

Ive, G. (1994), "A theory of ownership types applied to construction majors", *Construction Management and Economics*, 12(4): 349–364.

Javernick-Will, A. and Levitt, R. (2010), "Mobilizing institutional knowledge for international projects", *Journal of Construction Engineering and Management*, 136(4): 430–441.

Kim, D.Y., Kim, B. and Han, S.H. (2009), *Supporting Market Entry Decisions for Global Expansion Using Real Option + Scenario Planning Analysis*, Proceedings of the 2009 Construction Research Congress, Seattle, WA.

Korajczyk, R.A. (1996), "A measure of stock market integration for developed and emerging markets", *The World Bank Economic Review*, 10(2): 267–289.

Ling, F.Y.Y., Ibbs, C.W. and Chew, E.W. (2008), "Strategies adopted by international architectural, engineering, and construction firms in southeast Asia", *Journal of Professional Issues in Engineering Education and Practice*, 134(3): 248–256.

Ling, F.Y.Y., Ibbs, C.W. and Cuervo, J.C. (2005), "Entry and business strategies used by international architectural, engineering and construction firms in China", *Construction Management and Economics*, 23(5): 509–520.

Luo, Y. (2001), "Determinants of entry in an emerging economy: a multilevel approach", *Journal of Management Studies*, 38(3): 443–472.

Mahalingam, A. and Levitt, R.E. (2007), "Institutional theory as a framework for analyzing conflicts on global projects", *Journal of Construction Engineering and Management*, 133(7): 517–528.

Mody, A. (2004), "What is an emerging market?" *International Monetary Fund Working Paper 04/177*, online, available at: http://cdi.mecon.gov.ar/biblio/docelec/fmi/wp/wp04177.pdf.

Moreno, D. and Olmeda, I. (2007), "Is the predictability of emerging and developed stock markets really exploitable?" *European Journal of Operational Research*, 182(1): 436–454.

Nachum, L., Jones, G.G. and Dunning, J.H. (2001), "The international competitiveness of UK and its multinational enterprises", *Structural Change and Economic Dynamic*, 12(3): 277–294.

Nakata, C. and Sivakumar, K. (1997), "Emerging market conditions and their impact on first mover advantages: an integrative review", *International Marketing Review*, 14(6): 461–485.

Ngowi, A.B., Pienaar, E., Talukhaba, A. and Mbachu. J. (2005), "The globalisation of the construction industry – a review", *Building and Environment*, 40(1): 135–141.

Ofori, G. (2003), "Frameworks for analysing international construction", *Construction Management and Economics*, 21(4): 379–391.

Porter, M.E. (1990), *The Competitive Advantage of Nations*, Macmillan, London.

Rugman, A.M. and Hodgetts, R.M. (1995), *International Business: A Strategic Management Approach*, McGraw-Hill, New York.

Salomons, R. and Grootveld, H. (2003), "The equity risk premium: emerging vs. developed markets", *Emerging Markets Review*, 4(2): 121–144.

Seymour, H. (1987), *The Multinational Construction Industry*, Croom Helm, London.

United Nations (2005), *World Investment Report 2005, Transnational Corporations and the Internationalization of R&D*, United Nations, New York, online, available at: http://unctad.org/ en/ docs/wir2005_en.pdf.

Yeniyurt, S., Tamer Cavusgil, S. and Hult, G.T.M. (2005), "A global market advantage framework: the role of global market knowledge competencies", *International Business Review*, 14(1): 1–19.

9 Project delivery and financing: conventional and alternative methods

Michael J. Garvin and Ashwin Mahalingam

Introduction

Over roughly the last twenty-five years, the topic of project delivery and finance has received significant attention in the infrastructure community and the construction industry. In the United States, the public sector's mandated use of a delivery system where the activities of design, construction and operations were segregated, or independent, from one another has shifted toward a more open framework where other alternatives are now permitted and implemented. Globally, similar changes have occurred. The rationale behind these transitions is discussed subsequently. Further, this field remains in flux – a consequence, in part, of the search for ways to better align project stakeholder interests and objectives.

Many have offered definitions or descriptions of project delivery methods (e.g Miller *et al.* 2000; AIA-AGC 2004). We define a project delivery method as *an approach that organizes the tasks necessary for the development and/or operation of a constructed facility that provides a good or service.* We may consider the tasks as ones of: (1) asset creation and service provision such as designing, building, operating, and maintaining constructed facilities and (2) financing such as short-term or construction financing and long-term or take-out financing. We define financing in this context as *the acquisition of the capital necessary for the development, enhancement or modernization of a constructed facility.* Indeed, many project delivery methods are principally alternative combinations of the major tasks of design (D), build (B), operate (O), maintain (M), and finance (F).

The "world" of project delivery is full of acronyms and nuances. The term itself is often used interchangeably with "contracting strategy" or "procurement."[1] Often, contracting strategy is used when the perspective is the approach for the acquisition of a good or service, and a contract with an external organization is the means adopted. Similarly, the term procurement is used typically when a macro perspective of the project lifecycle is taken, and procurement is considered the mode of asset/service acquisition. The use of these terms is neither right nor wrong but rather one of viewpoint. For clarity, when we use the term "project delivery method," we will presume that the acquisition of goods or services will occur through contractual arrangements. Further, we will consider procurement as the process followed by an owner or a client for the acquisition of goods or services; in other words, the procedures used to select an external organization. The distinction between owner and client is made since the owner and the client are not necessarily one and the same in contemporary infrastructure projects; we shall have more to say on this subject later. Additionally, our focus is on infrastructure that facilitates public services.[2]

Our intent is to discuss this topic in a manner that first allows readers to understand the essence of project delivery and financing, so they may see its "roots." Once first-order concepts are presented, we explain conventional and alternative methods with an emphasis on alternative approaches, particularly why they are emerging and how they are implemented. To reinforce the latter, four cases from the United States and India provide concrete examples of the "why" and the "how." Subsequently, we diverge to explore emerging issues in this field, specifically the growing recognition that the efficient frontier of project delivery demands that we broaden our perspective of project arrangements beyond one that is based on the contractual framework. Finally, we conclude with several key points.

Fundamentals of project delivery

Constructed facilities require creation, operation, and maintenance, which involve a number of tasks throughout their lifecycles, most of which are interdependent. Figure 9.1 broadly depicts these activities. These are the tasks necessary to develop the asset so that it may provide its service. These tasks must also be financed and managed, and their nature and organization will influence the amount to be financed and the processes to be managed. Thus, they are certainly correlated with one another, and yet, multiple variations in task organization, finance, and management are quite feasible – as history has illustrated (Miller 2000 and Garvin 2007).

At a very basic level, a project delivery method is some combination of the tasks in Figure 9.1 coupled with where the responsibility for managing and financing the various tasks resides. For instance, if: (1) the task of detailed design is viewed as the major task of design (D); (2) the tasks of construction and commissioning are grouped as the major task of build (B); and (3) the responsibility for managing the D-B interface is done by the contracted service provider and the responsibility for finance rests with the owner/client; then we will likely delineate this method of delivery as design-build (D+B), as depicted in Figure 9.2. Certainly, the major task of design could also involve some conceptual design; this depends on the amount of design completed when an owner/client initiates procurement. *Effectively, any project delivery method, as we have defined the term, may be characterized at a fundamental level by examining the tasks required of the contracted service provider and identifying where the responsibility for financing the tasks and managing the task interfaces rest.*

Acronyms so prevalent in the construction industry such as Build-Operate-Transfer (BOT) or Design-Build-Finance-Operate (DBFO) are variants of the characterization method just described. Essentially, both methods require the contracted service provider to assume the major tasks of D+B+F+O+M for a specified period of time; the difference in practice between the two monikers is often a consequence of: (1) jurisdictional/regional perspectives and/or (2) a project's funding or revenue source. In some regions, DBFO (or DBFOM) is used to refer to the same approach that others call BOT. For instance, the US typically adopts DBFO or DBFOM whereas India prefers BOT. Alternatively, BOT may be used when the service provider relies on user fees or tariffs as the revenue source whereas DBFO may be used when the service provider is paid periodically by the owner/client.

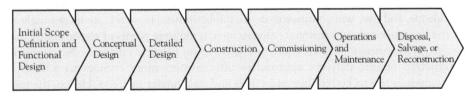

Figure 9.1 Constructed facility lifecycle tasks.

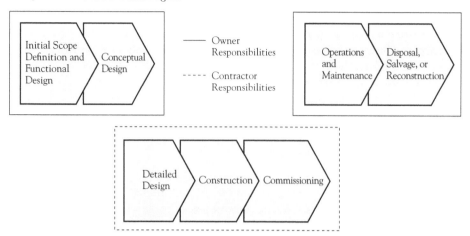

Figure 9.2 Task organization of design-build.

Another nuance, albeit a significant one, is whether or not the contracted party receives "ownership" rights, such as associated tax benefits – does the contracted party have the right to depreciate capital expenses? The answer depends on a project's location, as the taxation laws vary globally.

Conventional project delivery methods

General

Pinpointing the "conventional" method for project delivery depends on owner characteristics, industry sector and global location.[3] For instance, the public sector marketplace in Europe and Australia typically adopts design-build (DB) as the conventional approach whereas the same market in the United States and India utilizes design-bid-build (DBB). Regardless, conventional project delivery methods across the globe tend to have two defining characteristics:

*   The tasks of design and construction are segregated from the tasks of operations and maintenance. In other words, the process of developing a constructed facility is independent of the responsibility of service provision. In many cases, the owner/client of the constructed facility assumes the service provision role once the facility is commissioned.
*   The owner/client of the facility is responsible for long-term financing (and in many cases short-term or construction financing as well).

Conventional methods have played a significant role in the development of infrastructure worldwide, and they will continue to do so. Public owners, however, are increasingly aware of the limitations of conventional delivery such as multiple points of responsibility, lack of design and construction integration, and little to no lifecycle emphasis. Not surprisingly, alternative project delivery approaches offer owners more choices to address such shortcomings and to better align objectives and stakeholder interests. Hence, the use of a range of delivery options is becoming more common. In effect, the method of delivery has become a decision variable, so owners must determine what method to use and when.

Shift toward alternative project delivery methods

Why are public owners using alternative delivery methods, aside from the desire for greater choice? The reasons are many, and they are often correlated. Frequently, the shift is a matter of public policy. Oftentimes, the move is for pragmatic reasons such as greater emphasis on lifecycle or whole-of-life value. A summary of a few of the principal factors influencing this transition follows:

- **New public service paradigm** – many authors have noted a change in the role of government in the delivery of public services. Over roughly the last quarter century, governments, particularly in industrialized nations, have been under increasing pressure to become more efficient and "business-like." Some have labeled this trend as the "corporatization" of government (Froud 2003). Essentially, governments become less involved in actual service provision and more involved in the management of it. Consequently, outsourcing and contracting become more prevalent. The Private Finance Initiative in the United Kingdom is probably the most well-known example of such a transition. In fact, a survey of a dozen national governments across the globe in the late 1990s indicated that a significant majority of the respondents expected "that the most successful government structure in 2010 will be one in which government focuses on policy and project/supplier management, allowing the private sector to deliver most traditional public services" (Economist Intelligence Unit and Andersen Consulting 1999).
- **Budgetary benefits** – coupled with the expected efficiencies of the new public service paradigm are the budgetary advantages that alternative delivery methods can bring. Foremost, governments do not need budgetary resources available in the present if they have transferred the responsibility for financing a project to a contracted private sector service provider. This situation obviously is true in projects that secure the financing exclusively with user fee charges, but it is also true in cases where a government periodically pays service fees to the contracted party. In both situations, the government has not increased its debt burden, and in the latter it has amortized its budgetary expenditures, i.e. spread them out over time. Critics point out, often correctly, that the transfer of financing responsibility to the private sector (or "off-balance-sheet" financing) is frequently done to reduce budget deficits or improve debt capacity. Alone, this is an increasingly difficult rationale to justify an alternative delivery approach.
- **Bundling/lifecycle advantages** – the integration of the major lifecycle tasks (or the bundling of constructed facility development with service provision) has distinct benefits. Foremost, it places the responsibility for and the risks of lifecycle performance with a single party and more often than not this party must comply with contractual technical and service standards. In theory, a party that is responsible for a constructed facility over its lifecycle will give greater attention to constructability, maintainability and operational ease. Further, if the contracted party fails to comply with contractual requirements, then the owner (and potentially other stakeholders) often has recourse. Alternatively, a contract may be structured to incentivize or reward good performance. The focus from asset development to lifecycle performance naturally increases the attention paid to the service that a constructed facility enables as opposed to the descriptive or technical features of the asset itself.
- **Private financing** – in some regions of the world, the creditworthiness of private sector service providers exceeds that of the government. In such cases, privately arranged financing is more efficient and cost effective.

- **Government capacity** – in certain circumstances, governments may not possess the institutional maturity to manage complex infrastructure projects, so private entities are better positioned to manage development and operation. In such instances, however, a government should likely supplement its staff with experienced consultants in agency roles.

Alternative project delivery methods

General

Alternative project delivery methods – often grouped or classified as "Public Private Partnerships" or PPPs – can offer advantages over conventional approaches in asset creation and utilization. In a conventional system, the focus is primarily on asset creation, and most of the project risks are borne by the owner/client. In alternative project delivery arrangements, the contracted entity is not only in charge of creating the asset but is also given at least partial responsibility for discharging the service specified by the asset and, therefore, may bear additional risks with regards to operating, maintaining and assessing the demands for the service. Further, this entity may also be exposed to external risks such as political capture and protests by special interest groups.

A spectrum of alternative project delivery approaches exists that progressively transfers greater risk and responsibility to the private service provider. An abstract view of this spectrum is shown in Figure 9.3. This spectrum should be viewed as a continuum, with many possible arrangements. At a very simple level, an 'Operations and Maintenance' (O+M) contract can be awarded to a service provider to operate and maintain an existing asset, such as a water distribution system. The service provider in this case, normally a private organization, may lease the existing infrastructure from the government, treat the water and either supply wholesale water to the government or distribute water to retail customers. The service provider would also maintain the infrastructure for a period of time and bear the risk of infrastructure disruptions or failures during the contract period.[4]

At a slightly more complex level, a single entity can design, build and operate an infrastructure asset (D+B+O+M). In such cases, this entity bears the design, construction and operations risks. Many specific arrangements exist that feature varied permutations and combinations of roles and risk sharing. On the one hand, the ownership of the asset resides with the public sector, while a private entity designs, builds, leases and operates the asset for the public entity. Alternatively, the Build-Operate-Transfer (BOT) mode could be used

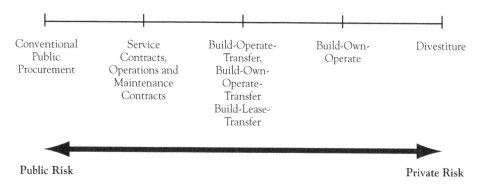

Figure 9.3 Continuum of project/service delivery methods.

where a private entity designs, builds, finances and operates an infrastructure asset (D+B+F+O+M); retains certain asset ownership and decision-making rights; and at a prespecified time transfers "complete" ownership of this asset back to the public client. The public entity can then decide how best to proceed with operating the asset. Finally, the responsibility of asset creation and management could be completely transferred or privatized, with a private service provider being given complete responsibility to build and operate the asset with no ownership or transfer back to the government; in this case, the government will often regulate the private enterprise through a commission or some other arrangement.

In all of these arrangements, parties to the project must agree on a method of payment, the duration for which the arrangement will exist, and the means by which performance will be measured. Payment may either be made directly by the owner/client to the service provider on a periodic basis (typically based on the availability of the infrastructure and hence known as an "availability payment") or can be collected directly by the service provider through levying user fees or tariffs. Owners/clients of such projects usually place greater emphasis on the monitoring of outcomes and the assessment of private operator performance with less specification of design and construction requirements to afford the private entity greater flexibility in devising asset development and service provision strategies.

Contracts where a private entity partially takes over ownership of an asset and later transfers it back to the responsible public agency are often called "concessions" or "concession agreements." Concession durations vary depending on the sector but are often awarded for several decades.

Structuring and financing alternative project delivery arrangements

Very often, a group of organizations with complementary expertise in developing and managing an asset join together and form a Special Purpose Vehicle (SPV) – an independent company solely responsible for a particular project. The SPV then injects a certain amount of equity into the project through contributions from the participating firms in proportions that indicate the level of involvement and ownership that a firm desires. Debt financing is also raised in order to cover project capital costs. This debt financing can be raised in several ways. In some cases, participating companies could use the strength of their balance sheets to mobilize loans from financial institutions. Alternatively, the SPV must raise debt purely based on the strength of the project. The latter approach is often known as non-recourse finance or project finance, wherein the lenders have recourse only to the assets of the SPV in case of project default. The lenders cannot apportion and do not have recourse to the assets or the balance sheets of the parent companies that form the SPV. Although this reduces some of the risks to the parent organizations, project financing is often more expensive and may take longer to arrange, incurring greater transactions costs. Regardless, many PPPs use the non-recourse SPV structure.[5]

Debt arranged by private entities is typically raised through loan instruments provided by commercial banks or through bonds sold in the primary market (which can subsequently be traded on the secondary market). Debt arranged by the public sector is often raised through general obligation or revenue bonds, where the former is secured by a public entity's tax-base and the latter is secured by system-wide revenues (such as those collected across a public water supply system). Debt may also have a hierarchical structure. Senior debt is normally the most common form of debt, but organizations may also raise mezzanine or subordinate debt, wherein repayments are made subject to payments being made on senior debt. Subordinate debt often functions in a manner analogous to a quasi-equity investment.

In a typical PPP with an SPV structure, the revenue collected by the private sponsor will flow to pay for the items shown in the order listed:

- Operations and maintenance expenses
- Senior debt
- Reserve funds, such as debt service or operations & maintenance reserve accounts
- Subordinate debt

Any remaining funds are then paid to equity investors or potentially retained by the SPV for future dividends or uses. The order of these items can vary from project to project, but the principle of distributing funds in a sequence that safeguards sustainment of the project and compensates debt and equity holders according to their cost of capital prevails.

The modes of financing may also depend on the structure of an SPV. In most cases, a government agency grants a concession to an SPV to create and maintain an asset and to deliver a service. In some cases, government agencies might themselves be a partner in the SPV, with a shareholder position and an equity contribution.[6] In such cases, a public agency would both award and partially execute the concession agreement.

Project stakeholders

Projects undertaken through the complex delivery arrangements described here often involve interactions between multiple stakeholders. The key stakeholders are illustrated in Figure 9.4 and described subsequently.

Owners/clients are often public sector or governmental organizations that are legally responsible for service delivery. They may use a careful process to determine the project delivery approach. If a decision is made to undertake projects through PPPs, owners/clients must then procure the services of an appropriate concessionaire and award the project to this entity.

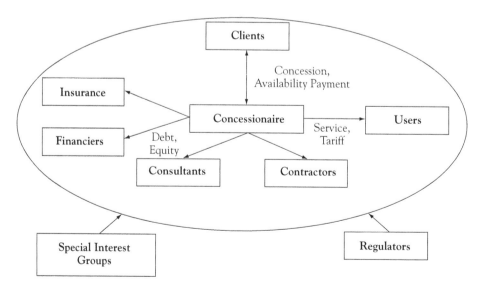

Figure 9.4 Typical PPP project stakeholders and structure.

Concessionaires are organizations that obtain concessions from owners/clients to create and maintain assets and deliver infrastructure services, usually for a specified period of time. Concessionaires can be formed from multiple firms and can be responsible for financing, designing, building and/or operating infrastructure services. They may function as the client for other stakeholders such as consultants, designers and operators. They may also receive certain asset ownership rights.

Citizens and special interest groups are the actual users of the infrastructure built and the services delivered. These groups play an important role in ensuring that projects are appropriately selected and delivered, and that owners/clients and/or concessionaires do not benefit at their expense. Citizen support is normally vital in such projects.

Financiers provide grant, debt and in some cases equity financing for projects. Finance can be raised from institutional investors, multilateral organizations, governments or private investors. Financiers assess the risk profile of a project in order to determine the amount and cost of financing that they are prepared to offer.

Regulators mediate between public and private entities and ensure that contractual provisions such as tariffs are set equitably bearing in mind the interests of the citizens. Regulators can also help guard against collusion and monopolistic practices within a sector.

Consultants provide the data and the analysis that both the owner/client and the concessionaire require to make decisions on the award and execution of the project. Consultants undertake technical and engineering studies as well as economic and financial studies that are vital to performing a transaction advisory role.

Designers, builders and operators are usually contracted by the concessionaire to help design, build, and operate and maintain the asset that is being developed. In many cases, concessionaires themselves have such expertise and often pass the responsibility of designing, building or maintaining to a representative or subsidiary from within the consortium. These organizations, along with technical consultants, are primarily responsible for the engineering and technical work involved in creating and operating an asset.

Finally, there are several other players such as insurance agencies who play a key role on such projects.

Table 9.1 Overview of case projects

Project	Location	Sector	Delivery Method
I-595 Express Lanes	Florida, USA	Highways	DBFOM to build reversible lanes in existing corridor with revenue source from government availability payments
Capital Beltway HOT Lanes	Virginia, USA	Highways	DBFOM to build HOV/HOT lanes in existing corridor with revenue source from tolls collected from variably priced HOT lanes
Alandur Sewerage Project	Tamil Nadu, India	Wastwater	DB to build sewerage network and BOT to build and operate treatment plant; revenue sources from user deposits for connections to the network and government payments for wastewater processed at plant
Tirupur Water Supply Project	Tamil Nadu, India	Water	BOT to build treatment plant and distribution network; revenue source from differential fees paid per kilo liter of water consumed by industrial and residential users

Case examples

We now present a series of case studies of projects delivered by alternative method to illustrate how they are structured and their inherent risks and outcomes. Table 9.1 provides an overview of the projects.

I-595 Express Lanes

Background and need

Opened in 1989, the I-595 Corridor is a key east-west highway link in Broward County in southern Florida, connecting I-75 and the Sawgrass Expressway in the west with Florida's Turnpike, I-95 and U.S. 1 in the east. Also known as the Port Everglades Expressway, the 13-mile long corridor facilitates access to major destinations such as the Fort Lauderdale-Hollywood International Airport and Port Everglades, and it is pivotal in southeast Florida's transportation network.

The I-595 Corridor was originally built on a twenty-year projection of traffic demand, with a planned capacity of 120,000 vehicles per day. The six-lane corridor, however, has experienced surges in traffic volumes, causing significant congestion. Daily traffic soared to 180,000 vehicles per day in the first decade, with traffic studies predicting a daily traffic demand of 300,000 vehicles by 2034. The Florida Department of Transportation (FDOT) recognized the need to improve the corridor to reduce congestion and improve safety conditions, so it completed an I-95/I-595 Master Plan in 2003. The Master Plan study proposed a Locally Preferred Alternative (LPA) for developing the I-595 Corridor; a key feature of the LPA was the construction of reversible express lanes.

Structuring the project

In 2004, FDOT began the I-595 Project Development and Environment (PD&E) study to further develop the LPA identified. The PD&E study solicited public opinion about the proposed alternatives through a Public Involvement Program (PIP) where public workshops were held to inform the public about the planned improvements and receive public feedback. The PIP was instrumental in disseminating project information to various groups such as government and regulatory agencies, local municipalities, county officials, legislators and communities. After incorporating the collective feedback, the I-595 PD&E Study was completed in March 2006, and the Federal Highway Administration (FHWA) approved the design concept for the preferred alternative in July, which cleared the way for subsequent necessary environmental approvals and potential federal funding. In June 2006, FDOT issued a notice to proceed to its selected design consultant to create an Indicative Preliminary Design (IPD) to further refine the scope and functional design elements.

FDOT originally intended to complete the needed improvements by issuing separate construction contracts over a period of years and funding the work on a pay-as-you-go basis. FDOT, however, began to consider other approaches and believed it could deliver the improvements perhaps fifteen years earlier if it bundled the work into a single contract as long as an appropriate financing strategy was identified.[7] FDOT, however, was concerned about two principal issues. First, a single contract would require a bond issue larger than any in FDOT's history, and it would absorb a substantial portion of its overall debt capacity.

Second, the project's scale exceeded any of its past projects, so FDOT was uneasy about completion risks, capital cost overruns and the concomitant operations and maintenance costs. Consequently, FDOT began exploring two alternative delivery options to reduce both the potential financial obligations and project risks – a design, build and finance (DBF) scheme and a design, build, finance, operate and maintain (DBFOM) scheme.

In 2007, FDOT held an industry forum where interested parties were invited to learn about the project and the alternatives under consideration. Over 400 participants attended. Subsequently, FDOT met individually with teams or firms that showed an interest in the project. A significant outcome of these meetings was the preference that the private sector showed for a DBFOM scheme.

Development and implementation

Following the industry forum, FDOT initiated a formal Value for Money analysis in August, 2007 to compare the merits of a DBF strategy with a DBFOM approach. The analysis assumed that FDOT would retain the fees collected from the proposed tolled express lanes. So under a DBFOM arrangement the contractor would receive its revenue from periodic direct payments from the government as either shadow tolls or availability payments. The analysis indicated that a DBFOM scheme using availability payments was preferable to the DBF method.[8] FDOT also debated whether or not to collect the tolls itself or transfer this right along with the accompanying revenue risks to the contractor. Ultimately, FDOT decided to maintain the toll revenues and pursue an availability-based funding approach for several reasons to include: (1) ensuring that the private contractor focused on enhancing throughput, not on maximizing revenue and (2) deterring political or public resistance of transferring toll collection rights to a private entity.

FDOT chose to deliver the project using a DBFOM approach with availability payments over a thirty-five-year contract term. In October, 2007, FDOT issued an Request for Qualifications and subsequently received and evaluated six Statements of Qualifications. Four teams were shortlisted: ACS Dragados-Macquarie Partnership; Direct Connect Partners (a Skanska Infrastructure Development, Laing and Fluor joint venture); Express Access Team (a Babcock & Brown and Bilfinger-Berger BOT joint venture); and I-595 Development Partners (an OHL Concesiones, Goldman Sachs Global Infrastructure and Balfour Beatty Capital joint venture).

In December 2007, a draft Request for Proposals was issued to the short-listed teams. An iterative process of communication between FDOT and short-listed teams then followed where FDOT evaluated feedback from its potential bidders with respect to its procurement process, concession terms and technical requirements. During this period, Direct Connect Partners withdrew from the process. A final RFP was then issued in April 2008, with final, detailed proposals due in September. Concurrently, FDOT also pursued federal authorization for the use of private activity bonds (PABs) and a Transportation Infrastructure Finance and Innovation Act (TIFIA) loan.[9] The U.S. Department of Transportation granted the project a provisional allocation of up to $2 billion in PABs while the process to pre-approve a TIFIA loan for up to 33 percent of eligible costs was started.

Two days prior to the submission date, I-595 Development Partners informed FDOT that it would not submit a final proposal. Hence, on September 5, 2008, only two final proposals were received. The proposals were evaluated on technical merit and price. Ultimately, the ACSID team emerged as the selected contractor, submitting a maximum annual availability payment of $63,980,000 as compared to the Express Access Team's $144,497,830, which substantially offset the higher Express Access technical score.

The ACSID team's financial plan originally envisioned a combination of PABs in two tranches, a TIFIA loan, and nearly $154 million in equity to cover the anticipated capital expenses of $1.3 billion. The turmoil in the U.S. and global financial markets that erupted in the fall of 2008, however, made it apparent that this financial plan needed reconsideration. ACSID and FDOT were both concerned that the PAB market would not produce the capital needed or it would come at too great a cost. Consequently, the ACSID team restructured the financial plan, replacing the PABs with commercial bank loans from a consortium of twelve lenders. The financial close for the project was achieved in March 2009 and the contract was executed shortly thereafter. Anticipated sources of financing include: $781 million in commercial bank debt, $603 million TIFIA loan, $208 million in equity, and $232 million in FDOT qualifying funds.

Discussion

The I-595 case illustrates how a public agency, FDOT, identified an infrastructure need and then progressed through its project development and delivery decision process, ultimately choosing to solicit a concession arrangement with an availability-based payment mechanism. The approach adopted allowed FDOT to deliver the needed enhancements far sooner than it might have through conventional delivery. This case also effectively illustrates the complex, systematic and iterative nature of the project shaping process and the need to balance various stakeholder interests in a transparent and accountable fashion. The procurement itself weathered a substantial economic crisis since the concessionaire was able to alter its financial plan, switching its debt acquisition strategy from private activity bonds to commercial loans. Undoubtedly, the ability to make this switch was enhanced by the overall strength of the project's owner/client, business case, and structure.

Capital Beltway (I-495) HOT Lanes

Background and need

The Capital Beltway (I-495) was initially constructed in 1956 and completed in 1964. It serves as a perimeter highway circling Washington, D.C. In 1977, four additional lanes were added to the existing four lanes; this was its last major improvement. Originally designed to serve through traffic bypassing Washington, D.C., the primary use has shifted towards local traffic with more than 75 percent of the current travelers along the Virginia section of the Beltway beginning or ending their trips in Fairfax County. The Beltway totals 3 percent of the lane miles in Northern Virginia while carrying nearly 11 percent of all daily regional trips. Without improvements, future growth would lengthen periods of severe congestion.

Structuring the project

Realizing that the congestion issue along the Beltway required action, the Virginia Department of Transportation (VDOT) completed a Major Investment Study in 1994, concluding that highway improvements on the Beltway should promote High Occupancy Vehicle (HOV) and bus travel in the region to address the area's congestion problems. In

1998, VDOT and the Federal Highway Administration (FHWA) began an Environmental Impact Study (EIS) to examine various improvement alternatives.

During this period, the state of Virginia passed the Public-Private Transportation Act (PPTA) in 1995 that enabled state and local authorities to enter into agreements with the private sector to provide needed transportation infrastructure that could not be funded out of the state budget. PPTA is the legislative framework enabling VDOT to enter into agreements with private entities to construct, improve, maintain, and operate transportation facilities. The act allows for both solicited and unsolicited proposals. VDOT maintains the right to solicit competing proposals when an unsolicited proposal is received in order to promote competition and improve the value for money of the proposed project.

In 2002, FHWA approved the EIS that included several HOV lane alternatives for the Beltway. In the same year, VDOT received an unsolicited PPTA conceptual proposal from Fluor Daniel to develop, finance, design, and construct High Occupancy Toll (HOT) Lanes on the Capital Beltway. Although VDOT advertised for competing proposals, none were received. In the spring of 2003, VDOT submitted a grant application to FHWA to study HOT lanes and other "value pricing" applications in Northern Virginia; it also held several public input meetings to solicit input regarding HOV versus HOT lane alternatives. A strong majority of the public feedback supported the HOT lanes concept. Early in 2005, the state's Commonwealth Transportation Board selected the HOT lanes plan as the LPA. By 2006, FHWA gave its final approval of the HOT lanes plan. In September 2007, Capital Beltway Express LLC, a joint venture between Fluor and Transurban, and VDOT reached an agreement in principle for the design, construction, operations and maintenance of the Capital Beltway HOT Lanes. This comprehensive agreement was finalized on December 20, 2007. Under this agreement, VDOT will own and oversee the HOT lanes and the concessionaire will construct and operate them. The total length of the concession is eighty years – five years of construction and seventy-five years of operation.

The project would build fourteen miles of new HOT lanes (two in each direction) on I-495 between the Springfield Interchange and north of the Dulles Toll Road in Northern Virginia in the United States. Tolls for the HOT lanes will change according to traffic conditions, which will regulate demand for the lanes and keep them congestion free. The project will be electronically tolled using transponder technology. This project will also make a contribution to the Beltway's forty-five-year-old infrastructure, replacing more than fifty aging bridges and overpasses, upgrading ten interchanges and enhancing bike and pedestrian access.

Development and implementation

Construction began in the summer of 2008, and the HOT lanes are expected to be open for service in 2013. A key aspect of the project was the effort to gain public support. The promise of the HOT lanes is the trip reliability it will allow to both public transit and commuters along this highly congested corridor; both have access to traveling lanes that are expected to provide an average travel speed (55[THIN]m.p.h.) with the latter paying a toll for use if a vehicle has less than three travelers. Capital Beltway Express and VDOT made a concerted effort to assure the public of these anticipated benefits.

The total $1.93 billion in expected costs are being financed through:

- $587 million senior debt in private-activity bonds (PABs)
- $587 million in Transportation Infrastructure Finance and Innovation Act (TIFIA) loans

- $350 million in equity (Transurban 90 percent and Fluor 10 percent)
- $409 million of VDOT funds

The internal rate of return (IRR) is projected to be 13 percent when the road begins operation, and the concession includes a revenue-sharing agreement with VDOT where the Department will receive a portion of the gross revenue once certain levels of return are met on the project. VDOT's entitlement starts at 5 percent when IRR is over 12.98 percent, rising to 15 percent when IRR is over 14.5 percent, and 40 percent when the IRR exceeds 16 percent. Transurban will also receive 1 percent of the net asset value of the concession as a base management fee.

Discussion

This project provides some insight into how alternative project delivery approaches can be initiated by the private sector, and the complex financial structures that are common to such projects. The unsolicited proposal submitted was in effect an alternative to the HOV lane improvements along the Beltway that VDOT was contemplating at that time. After lengthy study and public input, VDOT and FHWA concluded that HOT lanes were preferred to address the congestion issue in the region. Such diligence is often necessary to ensure the validity of a non-conventional project delivery approach. The project includes a mixed financing plan with senior debt in the form of PABs, a TIFIA loan, sponsor equity, and a VDOT contribution. The agreement also includes a revenue-sharing arrangement between VDOT and the concessionaire; these arrangements are becoming more common in PPP projects as a means to provide the public sector a share of any significant "upside" profits. The project is certainly an example of a joint effort between the public and private sectors to identify and implement a needed transportation project.

Alandur sewerage project

Background and need

Alandur is a town in the south of India with a population of nearly 150,000. A relatively low cost of living combined with a proximity to several industrial and information technology units led to an increase in Alandur's population in the early 1990s. This exerted a strain on the existing urban infrastructure, forcing the municipal authorities to plan for an expansion. The current state of sewerage infrastructure was particularly problematic. In the early 1990s, no integrated sewerage system existed, and residents built individual septic tanks that often overflowed and resulted in health hazards. The municipality turned its attention towards building world class sewerage infrastructure in Alandur.

Structuring the project

An initial feasibility study estimated the cost of construction as INR 453.1 million (approximately USD 10 million). Alandur's revenue base was a fraction of this amount and it was clear that the municipality could not finance the project on its own. Alandur Municipality hired the Tamil Nadu Urban Development Fund (TNUDF) – a public sector organization focused on helping municipalities finance and develop urban infrastructure. TNUDF suggested splitting the project into two packages. The first package would involve the construction of an integrated sewerage network, would be financed through user deposits as well as debt, and would be procured through a traditional engineer-procure-construct (EPC) contract. The second package would involve the construction and operation of a sewage treatment plant (STP) through a build-operate-transfer (BOT) concession. A private operator would finance, build and operate the sewage treatment plant and receive regular payments directly from the municipality based on the amount of sewage that was processed, subject to certain performance standards. Furthermore, TNUDF recommended that both packages be awarded to the same firm in order to ensure that the construction of the sewerage network would align with the development of the sewage treatment plant. Figure 9.5 depicts the project's structure.

Development and implementation

Alandur's residents lacked confidence in their municipality's ability to implement infrastructure projects and were therefore reluctant to pay deposits for the sewerage network. The mayor spent a considerable amount of time in canvassing the project to the residents. He organized meetings at street corners, met with residents' welfare associations, extolled the project's benefits through a loudspeaker as he toured the neighborhoods, and made the financial details of the project transparent so that the citizens could understand the rationale behind the project costs and financing charges. This effort paid off and the value of the deposits collected exceeded expectations. As a result, the debt component and the resultant financial burden on the municipality decreased.

Interested firms were asked to submit technical proposals for the project. Fifteen firms submitted proposals that were evaluated. Three of these firms that were deemed to have competent technical know-how were then asked to submit financial bids on the cost required to build the sewerage network as well as the number of years for which the sewage plant would be run by the private operators. For the purposes of the bid, the private operators were asked to set up two STPs of 12 MLD (million liters per day) capacity. The price per MLD and

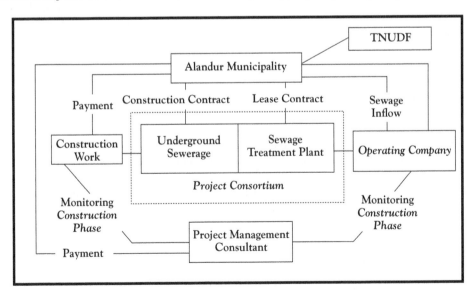

Figure 9.5 Project structure for Alundur sewerage project.

the minimum guaranteed flow were both fixed. A normalized combination of the construction cost and the proposed concession period were used to select the winning bidder.

Several hiccups and disputes occurred post-award. As per the contract, the municipality was tasked with providing connections from the boundary of a property to the sewage network. The municipality attempted to renegotiate this clause and asked the private operator to provide these connections, at a cost of INR 25 per meter – a price that was not appealing to the private contractor. Another private firm was then selected for this portion of the project.

Next, even as parts of the sewerage network and the pumping stations were built, municipal officials were slow to take them over citing a lack of capacity to manage these assets. A change in government then led to the replacement of project officials from the government's side. This new team caused delays while coming up to speed with the project.

TNUDF then withdrew from the project, even as the municipality questioned the readiness of the STP. Without TNUDF's presence, the municipality was unsure as to how to deal with the operator. When the STP was finally commissioned, the sewage inflow was much lower than expected. The municipality refused to pay for the minimum guaranteed amount of sewage and only relented when the private operator threatened to stop operations. Arbitration proceedings are currently underway as the operator seeks compensation for the delayed inflow of sewage into the system. However, the project was completed on time and is delivering its intended benefits to the residents of Alandur.

Discussion

This case highlights the challenges that alternative project delivery approaches face through their lifecycle. One of the key features of this case was the active role that the mayor played in order to engage and involve the municipal residents in the project. This led to a mitigation of potential conflicts. Also, Alandur municipality was not used to, and did not appreciate the complexity of, managing this PPP contract. TNUDF played a key role in bridging this competency gap and helping the municipality procure the project. However, their departure triggered a competency lapse which then led to disputes, arbitration and a degradation in service delivery. An understanding of the complexity that accompanies such projects and a willingness to work towards overcoming this complexity are therefore necessary prerequisites to undertake and sustain such approaches.

Tirupur water supply project

Background and need

Tirupur is located in the south of India and is home to a thriving textile industry. The textile industry is water intensive. With the growth in the Indian economy, the demand for textiles and consequently for water grew rapidly in the early 1990s. This, coupled with a lack of adequate rainfall in the corresponding time period, led to a situation where Tirupur Municipality was unable to provide a sufficient quantity of water to its residents and the nearby textile units. Tirupur municipality turned their attention towards finding a solution to this problem.

Structuring the project

Urged by the Tirupur Exporter's Association (TEA), the Government of Tamil Nadu (GoTN) instituted a company called the Tamil Nadu Corporation for Industrial Infrastructure

Development (TACID), and a project called the Tirupur Area Development Project (TADP). TACID would develop infrastructure in Tirupur with funds available through the TADP. TACID did not have the requisite competencies to undertake this project and partnered with Infrastructure Leasing and Financial Services (IL&FS) – a leading infrastructure consulting firm. Initial studies indicated that the costs of supplying water to Tirupur would be of the order of INR 4.5 billion (approximately USD 100 million) – an amount that was beyond the budgetary capacity of the GoTN. IL&FS therefore proposed a PPP approach. The GoTN accepted this proposal on the condition that water would be delivered at subsidized rates to the residents of Tirupur and its nearby villages.

GoTN, IL&FS, and TEA then formed a special purpose vehicle called the New Tirupur Area Development Company Limited (NTADCL) that would supply water to Tirupur's industries, to the town itself, to selected neighboring villages, and would provide a municipal sewerage system in the town of Tirupur. In order to make the project financially viable, NTADCL decided to implement a differential tariff system, wherein industrial consumers would pay higher rates and thereby subsidize residential consumers who would pay lower rates for water. This strategy did not sit well with the TEA initially, who were themselves part of the consortium and were faced with having to pay high rates for water. However, the project would not have been feasible without these differential tariffs and the textile industry was in dire need of water.

A concession agreement was signed between GoTN and NTADCL. NTADCL would extract and treat a maximum of 185 MLD of water from the nearby River Cauvery. Roughly half of this would be given to the residents of Tirupur (at a rate of INR 5 per kilo liter) and to the wayside villages (at a rate of INR 3.5 per kilo liter), while the other half would be given to Tirupur's industrial units (at a rate of INR 45 per kilo liter). NTADCL could earn a return on their investment of up to 20 percent, and the overall concession duration was thirty-three years from the award of the contract.

The project was financially structured with a debt to equity ratio of 1.5:1. Raising finances proved problematic since this was a first-of-its-kind project in India and potential lenders were unsure of the risks that the project entailed. The Industrial Development Bank of India finally agreed to lead a consortium of financiers who would provide debt to the project. They insisted however, that a water shortage fund be created to mitigate the risk of a lack of water in the river Cauvery. They also instituted a debt service recourse fund to protect the lenders against shortfalls in revenue. ILFS and GoTN also formed a joint venture company called the Tamil Nadu Water Investment Company (TWIC) that was a leading Equity investor in the project. Figure 9.6 depicts the structure of the project.

Development and implementation

The construction of the water supply system was divided into two parts, each of which was bid out separately. In addition, a third contract for the operations and maintenance of the entire system was also bid out, on a standard fixed-price basis. However, this was done for the sake of convenience and NTADCL's intention was to award all of these contracts to the same entity. These contracts were awarded on a typical engineer-procure-construct basis, based on the lowest bid that was received. A consortium headed by Mahindra Realty and Infrastructure Developers Ltd. were the winning bidders.

By and large the construction of the project was completed on time and within the expected budget. Work commenced in 2002 and was completed in 2005. As operations commenced, it appeared that the project was proceeding smoothly. However, by 2008 it was

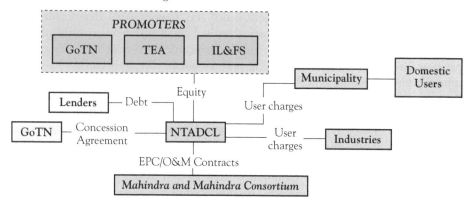

Figure 9.6 Project structure for Tirupur water supply project.

clear that water consumption by the industrial users was far less than expected. This was due to the economic recession that had lowered demand for textiles, heavy rains that had helped increase the ground water table, and stringent policing by the Tamil Nadu Pollution Control Board that hindered industrial activity. These factors combined to result in a revenue shortfall that was nearly 60 percent lower than had been predicted at the start of the project. NTADCL was therefore in grave danger of not being able to pay back the loans that they had borrowed to initiate construction.

NTADCL pursued many strategies to address this situation. They first lowered their prices in order to attract more customers. They also conducted and disseminated several scientific studies that showed the relative purity of NTADCL water over groundwater, and the consequent financial impact on the textile industry. However, these efforts were slow to bear fruit. In 2009, the GoTN infused nearly INR 80 million into the project in order to sustain it.

Discussion

This case provides an insight into some of the challenges faced when governing projects with alternate delivery mechanisms such as PPPs. The project's feasibility and sustainability rested on accurate demand and willingness-to-pay assessments that were made prior to the start of the project. In hindsight, these parameters were overestimated, and the resultant project revenues were insufficient to cover the project costs. This situation was exacerbated by factors external to the project such as the global recession which led to a decrease in textile output. Further, despite the fact that legislation existed to prevent the extraction of groundwater, this legislation was never enforced, highlighting the political nature of water supply projects. As a result, the concessionaire was hard pressed to continue delivering the service. On the other hand, the project was successful in that it did provide reliable, good quality drinking water to the residents of the municipality and the wayside villages. Despite failing on economic grounds, the project delivered valuable social benefits, justifying the government's decision to pump in funds to resuscitate the project.

Complexity of alternative project delivery arrangements

Clearly, the cases presented do not describe the entire range of available delivery methods, but they do provide examples that differ widely with respect to scope, sector, structure, risks, and outcomes. A common thread running through all of these cases is the complexity of the project arrangements. At least four kinds of complexities exist – organizational, procurement, risk, and financing – and each are briefly discussed.

Organizational complexity

All of the case study projects were characterized by a large number of stakeholders. The cases had a public client, private sponsors/contractors, users of the services provided, citizens affected by the projects, special interest groups, and financial institutions. Given the duration, size, and scope of such contracts, most of the stakeholder groups are often a coalition of many organizations. A project concessionaire is often composed of several organizations skilled at both developing and maintaining assets. The same is true of financial consortia. In addition, several technical and economic consultants typically play a role in shaping a project and public sector regulatory authorities help govern these arrangements. In order for such projects to be successful, the interfaces between each of these stakeholders need to be managed through the duration of asset creation and service delivery. In some cases, the objectives of these stakeholders may appear to be in conflict. For instance, a profit orientation on the part of private firms might seem to be at odds with service delivery to residents below a certain income level. Successfully managing such projects is therefore the art of using contractual and non-contractual strategies to ensure that such conflicts do not undermine the legitimacy of the project.

The chief challenge often lies in the fact that the roles that these stakeholders play in such projects differ considerably from their roles in more standardized project delivery approaches. Owners/clients, for instance, are often tasked more heavily with monitoring the performance of the private firms and are not as involved in assessing the processes that they follow. This represents a fundamental cultural shift in the way client-contractor relations are perceived. Groups that are slow in making this shift often need to deal with higher transactions costs when implementing such projects.

Procurement complexity

The I-595 Express Lanes and the Alandur sewerage system cases illustrated the use of complicated selection methods for a private contractor. Both employed complicated formulae and/or intricate selection algorithms/processes, which are the norm in alternative project delivery methods. A simple low-cost framework is unlikely to work. Various key decisions need to be made in the procurement process. Should the concession agreement be competitively awarded? Or should it be assigned to a qualified agency or organization that will then contract the actual tasks to be undertaken, as was the case in Tirupur? Should the government act purely as a client and as a regulator, or should they be part of the concessionaire organization? It is often very difficult to identify the optimal combination for a particular situation. When a combination is identified, contractual agreements need to be drafted that will: (1) safeguard the interests of the public and private sectors, as well as the broader stakeholder community; (2) define standards for construction and service delivery; and (3) hold over long time horizons. Given the complexities inherent in such projects, this is often a Herculean, if not impossible task. As classical economic theory predicts, most contracts must necessarily be incomplete.

Complex risk assessment and mitigation arrangements

Conventional contracts are characterized by a series of risks that encompass technical and legal project parameters. Alternative arrangements such as BOT face economic, political and socio-environmental risks, over and above those encountered in a traditional contract. Resistance from special interest groups, dips in demand due to changing economic circumstances, or an unwillingness to pay can ruin a project. A variety of risk mitigation tools need to be employed to hedge against such risks. Political Risk Insurance, for instance, can guard against political agency and expropriation. Providing for expenses and payments in the same currency can help hedge currency risks. Some risks can be managed by incorporating more sophisticated predictive models or state-of-the art engineering techniques, while others can only be addressed by taking on a diversification strategy. Finally, some risks that cannot be adequately addressed must be embraced.

Complex financial structures

As the cases indicate, the large gamut of risks that such projects face often leads to difficulties in financing such projects. Risk perceptions vary, and lenders may consider a project more risky than the sponsors. As a result lenders may price their loans at higher rates than project sponsors would expect to pay. Furthermore, financing is rarely done in isolation, but more often through a syndicate. Putting together an appropriate syndicate is a time-consuming task.

In addition to these challenges, arriving at an appropriate financial structure is a complicated endeavor. First, questions are posed on the relative proportion of debt and equity to be used. Next, the mode of financing used – corporate or project finance – the kind of debt that must be raised – senior or subordinate – and the sources of debt – institutions, bonds, multilaterals, etc. – must be determined. Finally, the project cost, the debt-equity ratio, the term of the concession, the interest rates on the loan taken, the return on investment expected by the concessionaire, and the project's revenues need to be carefully balanced. Arriving at a financial structure that provides an adequate return to investors, a viable set of tariffs for users, and hedges lenders' risks is a complex, iterative task.

Governance of complex infrastructure projects

Given that PPP projects are exceedingly complex, how can they best be managed? Certainly risks have to be assessed and allocated to the party best able to manage each risk, through contractual means. However, in conditions characterized by both uncertainty as well as complexity, complete determination of risks a priori is impossible. The impact of a risk, when it occurs is also difficult to predict. Therefore, contracts must necessarily be incomplete and mechanisms to govern such incomplete contracts must be devised.

Such "governability" must be introduced early on as the project is being "shaped." Three types of governance strategies exist – contract based governance, organizational structure based governance, and relationship based governance. Figure 9.7 illustrates the triangle of strategies that can be employed to ensure that such projects deliver benefits to prospective users over their entire lifetime. Each of these strategies is discussed briefly below.

Figure 9.7 The project governance triangle.

Contract-based governance

As we have seen, infrastructure projects that encompass both asset creation as well as service delivery encompass durations of several decades. Within this time there is the potential for significant exogenous economic, political, technological, social and environmental change. For instance, a recession might lower the ability of users to pay for a service. New environmental regulations that are more stringent in terms of emission controls might lead to the need to technologically revamp projects.

When such projects are first conceived and awarded, contracts are written that define the roles and responsibilities of the various parties involved. Significant analysis is done based on data collected on the ground as well as prevailing economic and social circumstances, to determine tariffs, concession lengths, and so on. However, this data cannot be relied upon to be completely accurate – it is often incomplete. Second, when exogenous change occurs, the need for a particular service or the ability of the community to sustain that service also changes. The original contractual and risk mitigation framework might no longer be sufficient. For instance, if the demand for a service decreases, the revenue accrued as a result of delivering the service will also decrease, leaving the project and the bearer of this risk in financial jeopardy.

Contractual agreements must therefore not be static and need to be living, flexible, evolving agreements (Orr 2005). As circumstances present themselves, there should be adequate scope to rewrite the contracts to redefine roles, responsibilities, covenants, and risks in a manner that is neither detrimental to the project nor to the various stakeholders. For instance, a response to a reduction in demand for a service could be to change the tariff charged for the service, or a change in the duration of the concession agreement. Contracts between the various project parties should be amenable to such modifications over the lifetime of the project.

Contracts can have triggers that lead to renegotiation. A fall in demand that is greater than a certain percentage, or a significant change in the wholesale price index can be some triggers that lead to renegotiation. Alternatively, contractual agreements can be periodically reviewed over fixed time periods in order to ensure that the project structure and risk allocation continues to meet the current demands of the project. In general contracts can be hierarchical in nature and can delineate circumstances under which renegotiation is possible and the rules and parties to such renegotiation.

Organizational-based governance

A misalignment of goals or incentives between various project stakeholders is often a key determinant of project conflict. Public sector clients have fundamentally different project objectives from private sector service providers. The former presumably is interested in providing high levels of service and enhancing their vote base, while the latter attempts to maximize their profits while meeting project objectives. Consultants at the planning stage are often incentivized to buffer their estimates in the hopes of bagging lucrative contracts in the development stage of the project (Levitt *et al.* 2009). Such differences in incentives often lead to poorly structured projects – and consequent project failure – as well as disputes and altercations between project parties.

While structuring a project, it is therefore imperative to ensure that organizational incentives are aligned and that all parties benefit if the project succeeds. This is best exemplified in the Tirupur water supply project, where the government held a share in the project vehicle. When the project encountered turbulence, all parties were able to hold an "around the table" discussion as opposed to an "across the table" discussion, since both the government and the private contractor were part of the same project vehicle, and shared the same project goals. There have been other instances of incorporating local residents into the contract by mandating that they monitor the performance of the private contractor and be responsible for ensuring good quality service delivery. This strategy ensures that residents are now parties to the delivery of the service and can ensure that their objectives are integrated with the private contractors' service delivery strategy.

In general, projects that incorporate strategies that align economic incentives across stakeholder groups are more likely to withstand turbulent times. Projects that do not do so often end in disputes, factionalism and one-upmanship. Service delivery is often compromised.

Relationship-based governance

Sophisticated legal and economic structuring of a project are necessary but not sufficient conditions for project success. Social and psychological factors such as the existence of trust between the citizens and the state play a key role in determining the success of complex projects. The Alandur sewerage case study is a fine example. Alandur's mayor embarked on an intensive campaign to showcase the project to his constituents. This campaign made the mayor accessible to his constituents and allowed them to participate in the process of project development. Further, the visibility and transparency of the campaign – an offshoot of the mayor's leadership credentials – led to an increase in the legitimacy of the project. It is therefore no surprise that Alandur's residents readily paid one-time deposits to finance the project. The process of project development was less transparent in other towns where similar projects were attempted, and without a project champion, the collection of user deposits was a failed effort.

The mayor's strategy in Alandur also reflects the use of a fair process and its role in project success. Studies show that when a process of development is perceived as being fair, community support and involvement are likely to be present irrespective of the end product being built. *Engagement, explanation* and *expectation clarity* are the three strategic arms of the fair process framework (Kim and Mauborgne 1997). The mayor engaged his electorate in discussing the project and encouraged suggestions from them. All aspects of the project including financial estimates were clearly explained, and the expectations from the government, the citizens and the project were clearly set out. The residents were thus satisfied on the merits of the project and supported the endeavor. Similarly, cooperative societies that deliver water supply to their constituents often succeed due to the fact that they are self-governed and all residents know and trust the project administrators.

Similarly, VDOT and its concessionaire worked throughout the project-shaping process of the Capital Beltway HOT Lanes project to both assess the willingness of users to utilize HOT lanes as well as to promote the benefits of the project to them. Ultimately, the public support gained was crucial to the project proceeding. While formal processes for public input and feedback are typical of a project of this nature in the United States, the ultimate success of this endeavor truly depends on whether the VDOT and its concessionaire deliver on the implicit "social" contract that developed. Public support hinged on the promised trip reliability of the HOT Lanes. Time will tell whether this promise is fulfilled or not.

PPP projects often face risks from citizen and special interest groups that can threaten to derail a project. Relational strategies that create an atmosphere of trust and teamwork can go a long way in sustaining complex project relations.

Conclusion

Project delivery methods are evolving, and they will continue to do so. We have attempted to "boil down" these arrangements to their essence – task responsibility, task interface management, and task financing – in order to allow the reader to comprehend the various permutations and acronyms that exist. These three elements are complemented by the revenue source and the ownership rights associated with any method.

The resurrection of alternative delivery methods provides owners a choice with respect to what role they want to take in a project and what services and risks they desire contractors to assume. A key benefit of concession arrangements or PPPs is the emphasis on service provision. This permits the owner/client to focus on the "ends" desired as opposed to the "means."

The cases illustrate the complexity of alternative approaches and suggest that a more comprehensive perspective of projects is needed. Such projects are more than contractual agreements. They are long-term arrangements where the attention given to organizational matters and relationship building is increasingly significant and challenging.

Final take-aways

1. At its essence, a project delivery method is an approach that contractually organizes the tasks of asset creation, service provision and/or financing.
2. Alternative delivery methods typically bundle the requirements of asset creation, service provision and financing into a single contract between an owner/client and a consortium of private enterprises; a principal, and often compelling, argument for these methods is their emphasis on lifecycle service and value.
3. A key element of any project, but particularly those delivered via non-conventional methods, is identifying a reliable source of revenue; while PPPs are often heralded as mechanisms for "additional" financing of projects, the amount and cost of potential financing are inextricably linked to the strength of the revenue source.
4. Alternative delivery methods are quite complicated since they frequently involve a diverse set of stakeholders, feature intricate arrangements and span decades; not surprisingly, project identification, procurement, contracting, and management are far more complex.

5. Prediction of the outcomes that such projects will encounter over their lifespans is virtually impossible; accordingly, such projects are best regarded as incomplete contracts that have to be monitored and governed on a continual basis.
6. The contemporary global maturity with non-conventional delivery methods varies and continues to evolve. Improving the performance of projects delivered in such fashions will likely depend on advancing how such projects are governed by flexible contracts, among involved organizations, and through personal/organizational relationships.

Notes

1   The use of 'contracting strategy' is likely related to literature in areas such as the theory of the firm whereas the use of 'procurement' is likely related to literature in areas such as facilities management or theories of purchasing/buying.
2   See Grout and Stevens (2003) for a more detailed discourse on public services.
3   The term "conventional" was adopted as opposed to "traditional," which may be more common; if the shadow of the past is considered, what may be non-traditional in a contemporary context actually has a longer history when the timeframe is centuries as opposed to decades.
4   The extent of this risk is typically defined in the contract or negotiated during the procurement process.
5   Yescombe (2007) and others discuss this issue as well as other issues of project finance.
6   This situation is not common in the United States and many European nations.
7   Additional advantages included cost efficiencies from a single mobilization effort as opposed to multiple mobilizations, less disruption to existing traffic through better traffic management, and economies of scale in service fees and materials purchasing.
8   The analysis between DBF and DBFOM using shadow tolls was far less indicative of one being a better alternative than the other.
9   PABs are bonds issued by or on behalf of a governmental entity for use by a private party whereas TIFIA is a U.S. federal credit program for eligible transportation projects of national or regional significance under which the U.S. Department of Transportation (DOT) may provide three forms of credit assistance – secured (direct) loans, loan guarantees, and standby lines of credit. TIFIA loans often act as mezzanine financing to bonds or commercial loans.

Recommended reading

Grimsey, D. and Lewis, M.K. (2004), *Public Private Partnerships: The Worldwide Revolution in Infrastructure Provision and Project Finance*, Edward Elgar Publishing: Northampton, MA.
Savas, E.S. (2000), *Privatization and Public-Private Partnerships*, Seven Bridges Press: New York.

References

AIA-AGC (American Institute of Architects and The Associated General Contractors of America) (2004), *Primer on Project Delivery*, AIA & AGC: Washington, D.C./Alexandria, VA.
Economist Intelligence Unit and Andersen Consulting (1999), *Vision 2010: Forging Tomorrow's Public-Private Partnerships*, EIU & AC: New York.
Froud, J. (2003), "The Private Finance Initiative: risk, uncertainty, and the state," *Accounting, Organizations and Society*, 28(6): 567–589.
Garvin, M.J. (2007), *America's Infrastructure Strategy: Drawing on History to Guide the Future*, KPMG LLP and Stanford University, Washington, D.C.
Grout, P.A. and Stevens, M. (2003), "The assessment: financing and managing public services," *Oxford Review of Economic Policy*, 19(2): 215–234.
Kim, W.C. and Mauborgne, R. (1997), "Fair process: managing in the knowledge economy," *Harvard Business Review*, 75(4): 65–75.

Levitt, R.E., Henisz, W.J., and Settel, D. (2009), "Defining and mitigating the governance challenges of infrastructure project development and delivery," in P.S. Chinowsky and R.E. Levitt (eds.), *Proceedings of the 4th Specialty Conference on Leadership & Management in Construction, November 5–7, South Lake Tahoe, CA*, CD-ROM.

Miller, J.B. (2000), *Principles of Public and Private Infrastructure Delivery*, Kluwer Academic Publishers: Boston, MA.

——, Garvin, M.J., Ibbs, C.W., and Mahoney, S. (2000), "Toward a new paradigm: simultaneous use of multiple project delivery methods," *Journal of Management in Engineering*, 16(3): 58–67.

Orr, R. (2005), *Proceedings of the First General Counsels Roundtable*, Collaboratory for Research on Global Projects, Working Paper Series, Stanford, CA.

Yescombe, E.R. (2007), *Public-Private Partnerships: Principles of Policy and Finance*, Butterworth-Heinemann: Boston, MA.

10 Strategic safety management

Matthew R. Hallowell and Jimmie Hinze

Introduction

To improve the quality of life within society, construction organizations build residences, commercial buildings, industrial projects, roads and bridges, and many other types of projects. While there are vast differences between the many types of projects, there is one common resource that binds them all: construction workers. Therefore, the safety and well-being of the workforce will directly impact the construction process and, ultimately, project and organizational success.

Fortunately, the safety performance of the construction industry has improved considerably over the past two decades. In fact, the Occupational Safety and Health Administration (OSHA) recordable injury rate (i.e., the number of medical treatment cases per 200,000 hours of worker exposure) has declined from 13.10 in 1992 to 5.40 in 2007. Although there has been obvious improvement, the construction industry consistently accounts for the highest fatality rate of all major service industries. Despite the fact that the construction industry only accounts for 7 or 8 percent of the US industrial workforce, the industry consistently accounts for over 20 percent of all occupational fatalities (BLS 2009; CPWR 2008). In addition, over 1,500 OSHA recordable injuries and 50,000 first aid injuries occur every working day. Clearly, the injury rate is at an unacceptable level and there is obvious room for improvement.

Many inherent characteristics of the construction industry contribute to the relatively high injury and illness rate including dynamic work environments, industry fragmentation, multiplicity of operations, proximity of multiple crews, and industry culture (Fredricks *et al.* 2005). Each of these characteristics contributes to unforeseen and unfamiliar hazards or the unsafe behavior of workers. It has been argued that companies do not need to be overly concerned about safety as the federal safety regulations promulgated by OSHA already provide guidance for the safe delivery of projects. Such arguments are made without the recognition that the OSHA regulations are minimum standards and compliance with these regulations alone will not result in world-class safety performance.

Excellence in safety performance requires a culture and climate that are centered on the well-being of the workforce. Construction organizations can promote a culture of safety by implementing a comprehensive safety management program, allocating sufficient resources for safety-related activities, ensuring the active participation of upper-level managers in regular safety management efforts, and integrating safety into all aspects of organizational management (Neal *et al.* 2000).

This chapter will describe the role of safety in business performance, offer a framework for effective safety management, describe the specific management activities that promote a culture of safety throughout the organization, and discuss the roles and responsibilities of upper management. The chapter concludes with a brief overview of emerging safety enhancement strategies that focus on the integration of safety into the lifecycle of a construction project.

Impact of injuries on business performance

While the statistics on construction worker injuries are reminders of the substantial amount of suffering that occurs on a daily basis as a result of worker injuries, it is important to understand how injuries impact business performance as well. In recent years, the relationship between safety and organizational performance has become a topic of great interest. Following the Occupational Safety and Health Act of 1970, employers have been required to provide their employees with a safe and healthful workplace. As a result, employers began to invest in personal protective equipment, signage, barriers, and other aspects of safety management to avoid citations and to preserve a positive reputation within the industry. Many organizations viewed safety management as a non-value-added business expense that was required only to meet legal obligations. Recently, however, researchers have established a positive relationship between investment in safety and organizational performance.

When a worker is injured, the organization realizes both direct and indirect costs. Direct costs are defined as expenditures that are covered by the firm's workers' compensation insurance carrier while indirect costs of injuries are broadly defined as expenditures that are not associated directly with the treatment of the injury itself, but which are incurred as a result of the injury. Examples of direct costs include the treatment of injuries by a physician, diagnostic tests, medication, hospitalization, and rehabilitation. Alternatively, lost or reduced productivity due to the injury, costs incurred to transport the injured worker to the healthcare provider, damaged materials and equipment, injury investigation expenses, injury reporting expenses, and additional training costs are all examples of indirect costs. Indirect costs generally are not included in traditional project accounting methods despite the fact that they are paid in real dollars and are not recovered through insurance.

Indirect costs have been determined to be substantial. For example, it has been shown that the indirect costs of a typical OSHA recordable injury equate to 40 hours of lost production time and nearly 10 hours of administrative time. According to the Occupational Safety and Health Administration (OSHA), the indirect to direct cost ratio ranges from 1.1 to 4.5, depending on the severity of the injury. The published ratios have been reproduced in Table 10.1 below for reference. Although there are other methods of calculating indirect costs, the use of these ratios for estimating purposes is an accepted standard in the industry.

Table 10.1 Indirect to direct cost ratios for occupational injuries

Direct costs	Indirect to direct cost ratio
$0 - $2,999	4.5
$3,000 - $4,999	1.6
$5,000 - $9,999	1.2
$10,000 or more	1.1

When direct and indirect costs are combined, the true costs of construction injuries and illnesses are staggering. In 2004, the National Safety Council reported that there were 1,194 fatalities with an average cost of approximately $1,150,000 per fatality. The NSC also reported that the construction industry as a whole experienced 460,000 disabling injuries resulting in an estimated total cost of $15.64 billion (NSC 2006). Though alarming, these published costs do not account for the low severity injuries such as persistent pain and first aid cases that contribute greatly to the overall safety-related costs. When considered in the context of construction projects, researchers have found that the total cost associated with all construction accidents accounts for 7.9–15 percent of the cost of new, non-residential projects (Everett and Frank 1996). These findings were confirmed in a case study conducted by the Health and Safety Executive (HSE) that revealed that injuries accounted for 8.5 percent of the tender price for the project.

In addition to direct and indirect costs, contractors must maintain low workers' compensation premiums and achieve consistently low injury and illness rates to be competitive in open bidding environments. Workers' compensation insurance represents a significant expense to contractors and typically accounts for about 20 percent of the costs of labor on a construction project or roughly more than 5 percent of the costs of many construction projects, depending on the type of project.

Workers' compensation insurance rates vary greatly in the construction industry and are determined using the Experience Modification Rate (EMR). While the workers' compensation premiums are based on manual rates (essentially a stipulated percent of the wages) which are established in each state for each trade occupation, the EMR is a modifier of that manual rate. The EMR is used by insurance providers to gauge past costs of injuries and future projections of risk for an organization by comparing the past claims history of the company with similar companies in the construction industry.

The EMR has a significant impact on business performance for several reasons. First, the EMR often serves as a direct multiplier of the base insurance rate set by the National Council on Compensation Insurance (NCCI) when calculating insurance premiums. For example, an organization that has an EMR of 1.2 can expect to pay about 30 percent more for their worker's compensation insurance premiums than a competitor with an EMR of 0.9. Therefore, an organization with a relatively high EMR is in a competitive disadvantage during the bidding process, especially when competition for projects is high. The average construction organization will have an EMR of 1.0 while organizations with a good claims history should expect an EMR significantly less than one (e.g. 0.70) whereas an organization with a relatively poor claims history could have an EMR that is greater than 1.0. When the EMR is greater than 1.3 or 1.4, companies often fail because they cannot compete successfully in the open market.

The NCCI developed a complex formula for calculating the EMR that considers the frequency and severity of injuries, trends in safety performance, and types of injuries. Generally, the major influence on the EMR is the Loss Ratio ($claims/$insurance premiums). That is, the loss ratio represents the costs of claims paid by the insurance carrier for every dollar paid in premiums. Because the insurance carrier must pay for the salaries of insurance agents, home office staff, rent on office space, utilities, and so on, the loss ratio must be considerably below one for the insurance company to break even. Some experts have suggested that the insurance carrier will lose money on a client if the Loss Ratio exceeds about 0.60. Thus, the insurance company needs about 40 cents of every premium dollar to maintain its business operations. Understandably, when the Loss Ratio approaches or is greater than one, the EMR will rise. The EMR is also affected by the type of injury. For example, a company that incurs the majority of its claims from one big loss will be penalized less severely than a company that incurs the same cost of claims but with many small losses because frequent low severity injuries are seen as an indication that more severe injuries will occur in the future. Finally, the EMR is based upon trends in performance. The computation of the EMR typically consists of a review of a company's claims history for the three years prior to the expiration of the current policy (exclusive of the immediately preceding year), thus penalizing or rewarding an organization based on their past claims history. For example, if a company's current insurance policy expires in 2011, the review of the claims history will include 2007–2009 when establishing the new premium rate.

In addition to affecting insurance premiums, the EMR and recordable injury rates are used by many owner agencies in both the private and public sector to prequalify bidders. Owner agencies are beginning to recognize that safety performance is an indicator of other organizational attributes such as timely project delivery; excellent housekeeping; strong relations with reputable suppliers and subcontractors; and high employee morale. Owners also seek to contract with safe contractors to maintain or improve their reputation.

It is clear that investing in safety enhances organizational performance. Nevertheless, one may ask: "What is the optimal investment in safety management?" Though there is a wealth of information related to the costs of construction injuries and the relationship between safety performance and competitiveness, there are few studies that investigate optimal investment strategies for safety. Hinze (2000) describes investing in safety as a "game of probabilities." The hypothesis is that injury costs will be high when the level of emphasis on safety is low and vice versa. While investing in safety is critical to safety success, there is a point where additional investment yields diminishing returns and the cost-benefit ratio is greater than one. Practically, however, this level is rarely achieved. For guidance, safety investment strategies implemented by highly effective firms for their basic safety program elements is discussed later in this chapter.

Framework for effective safety programs

When designing a site-specific safety program, there are many elements to choose from. In fact, Rajendran (2006) identified over 100 distinct safety program elements. Since most construction organizations have limited resources to allocate to safety management, contractors are forced to select a subset of the available elements. Several authors have identified the essential elements of an effective safety program (e.g. Coble *et al.* 2000; Gibb *et al.* 1995; Hinze 2006; Hill 2004; Jaselskis *et al.* 1996; Liska 1993; Rajendran 2006). Collectively, these works highlight 13 independent elements of an effective construction safety and health program. To further distinguish among these 13 components, Hallowell and Gambatese (2009) ranked these elements by their relative ability to mitigate risk. The 13 essential safety program elements, in order of effectiveness, are identified and described in Table 10.2. In addition to the efforts described in Table 10.2, workers should expect to wear a hard hat and safety glasses at all times and be tied off when working at elevation without other fall protection.

Before novel approaches are implemented, it is important to support and implement those programs that have become well-entrenched on construction sites. While the elements described in Table 10.2 are not generally considered to be sufficient to establish world-class safety performance, they are the mainstay of what many construction workers will expect as a minimum.

Table 10.2 Descriptions of highly-effective construction safety program elements

Safety program element	Description
Upper-management support	Explicit acknowledgement from upper management that worker safety and health is a primary goal of the firm demonstrated by participation in regular safety meetings and safety committees, and sufficient funding for safety.
Subcontractor selection and management	Consideration of safety and health performance during the selection and management of subcontractors (e.g. prequalification and required compliance).
Employee involvement and evaluation	Including all employees in the formulation and execution of other safety elements and including participation and safe work behavior in evaluations.
Job hazard analyses (JHA)	Review and record activities associated with a construction process, highlighting potential hazardous exposures, and documenting safe work practices to prevent injury.
Project-specific training/ meetings	Establishing and communicating project-specific safety goals, plans, and policies before the construction phase of the project.
Frequent worksite inspections	Inspections performed internally by a contractor's safety manager, safety committee, representative of the contractor's insurance provider or by an OSHA consultant to identify uncontrolled hazardous exposures.
Safety manager on site	Employment of a safety and health professional (i.e. an individual with formal construction safety and health experience and/or education) whose primary responsibility is to perform and direct the implementation of safety and health program elements and serve as a safety resource for employees.
Substance abuse programs	Identification and prevention of substance abuse of the workforce (includes pre-employment screening, random testing and post-accident testing)
Safety and health committees	Committee, with the power to affect change and set policies, consisting of a diverse group including supervisors, laborers, representatives of key subcontractors, and owner representatives. May be formed with the sole purpose of addressing safety and health on the worksite.
S&H orientation/training	All new hires or transfers must receive orientation and participate in training sessions that have a specific focus on safe work practices and company safety policies.
Written safety and health plan	Development of a documented plan that identifies project-specific safety objectives, unique hazards, and methods for achieving success.
Record-keeping/ analyses	Regular reporting of the specifics of all accidents including information such as time, location, worksite conditions, and cause.
Emergency response planning	Creation of a plan that documents the company's policies and procedures in the case of a serious incident or catastrophe such as a fatality or an incident involving multiple serious injuries.

The 13 elements described in Table 10.2 are implemented by a majority of the leading construction organizations in the US with very few exceptions. Outstanding safety performance, however, is not simply the outcome of implementing a series of safety elements in a safety program. Rather, safety is the outgrowth of a mindset that is established on a construction project and within a construction company. In fact, a safety program will fail if all the safety programs are put in place but there is no deep-seated conviction that accompanies it. The safety program is simply a foundation. Achieving safety excellence requires that upper-level managers build this foundation by setting safety as a core value of the company, demonstrating commitment through active participation, and integrating safety management throughout all project management processes and in all decisions. Once safety has been infused throughout the organization and personnel at all levels share common safety goals, a culture of safety will prevail. The following section details several management strategies that are essential to the development of a strong safety culture.

Establishing a culture of safety

Safety culture is a topic that has received considerable attention in recent years since researchers (e.g. Mearns *et al.* 2003) found a positive correlation between safety culture and accident reduction. This occurs in an environment in which safety is a core value, not simply a priority which can change. As a core value, it is understood that worker well-being is a constant and ongoing objective that will not be compromised to meet some other objective.

In order to achieve an excellent culture of safety, all employees must share the beliefs that injuries and illnesses are not an acceptable part of the work, safety is held paramount over all other project outcomes, and that safety is a critical component of all decisions. Most importantly, employees must demonstrate these beliefs by acting accordingly and holding one another accountable for safe work behavior. Laborers, crew leaders, foremen, superintendents, project managers, executives, and even CEOs all play an essential role in *maintaining* a culture of safety. *Developing* a strong safety culture, however, must start with top executives and senior managers. These individuals shape the vision of the organization, set objectives, and have the power to fund safety-related initiatives. The following eight components build effectively on the foundation set by the safety program to promote a strong safety culture within an organization:

1. Demonstrated commitment of upper level managers through active participation and resource allocation.
2. Establishing safety as a core value and establishing clear, achievable objectives.
3. Resource allocation for safety-related activities.
4. Involvement of employees at all levels in safety management, planning and execution.
5. Evaluating employees based on active safety participation and safe work behavior.
6. Integration of safety with all other management functions.
7. Strong ethical standards.
8. Continuous improvement.

The benefit of focusing on the eight dimensions above is that they (1) are indicators of safety culture; (2) can be measured or observed; and (3) have been objectively studied by researchers and linked to improvements in safety performance (Tarrants 1980; Sawacha *et al.* 1999). The subsequent sections of this chapter will focus mainly on the best practices associated with these eight dimensions and the specific roles and responsibilities of upper-level managers.

Figure 10.1 is a model of the relationships among the dimensions of an effective safety culture.

The model of safety culture presented in this chapter has been created based upon a review of relevant literature and the experiences of the writers. This model depicts the essential elements of a strong safety culture and the relationships among them. As one can see, the regular injury prevention strategies (the 13 essential elements presented in Table 10.2) serve as the foundation of a safety culture which must be supported through sufficient funding from upper management and a formal safety knowledge management system. Additionally, worker involvement and employee incentives/disincentives must be included to ensure employee buy-in and to promote participation. Finally, continuous improvement programs that incorporate continuous safety audits must be implemented to safety related learning within the organization. All of these safety efforts should be designed to support the organization's safety mission. Outcomes include reduction in injuries, improved business performance, enhanced employee morale, and other positive effects.

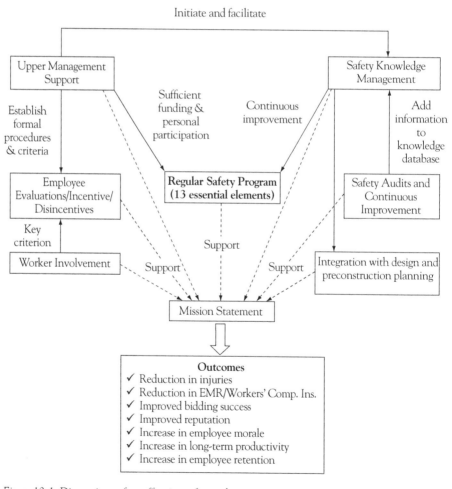

Figure 10.1 Dimensions of an effective safety culture.

Top management commitment to safety

An overwhelming majority of the studies that focused on highly effective safety management practices have concluded that the single most influential factor is the demonstrated commitment of upper-level managers to safety through resource allocation and personal participation in regular safety management activities (Hinze and Pannullo 1978; Jaselskis *et al.* 1996; Hallowell and Gambatese in press). Safe work practices can only be expected if company leaders demonstrate and enforce safe work practices as upper-level managers establish the link between organizational goals and practice. For example, on a project walk-through the president of a company can make a tremendous impression if time is taken to provide a friendly but firm admonishment to a worker for not wearing safety eyewear or if time is taken to praise a worker for being tied off when working at elevation. Of course, it is also imperative that the company president *always* wear the appropriate personnel protective equipment on such project tours.

In order for managers to be truly effective they must integrate safety with all aspects of the management process (Rowlinson 2000). This means that safety should be an inherent topic on the agenda of all meetings, considered during the selection of alternative means and methods of construction, and set as a key criterion for selecting subcontractors and suppliers.

Stated organizational objectives

There are many ways that top management can demonstrate its commitment to safety. One obvious way is through the mission statement. If profitability is the only issue addressed in the mission statement, it will quickly become clear to others that safety "plays second fiddle" to company profits. It is recognized that a company must be profitable in order to survive, but the well-being of the construction workers should never be ignored. A well-crafted mission statement posted prominently on the company's website is an effective way to promote and communicate safety goals.

Several model organizations include safety as a part of their organization's mission statements and/or published business strategy such as Skanska AB, McCarthy Building Company, and J.B. Henderson Construction Co. These organizations not only include safety as a core business value but publish this value on a prominent location on their website with clear visibility for all of their employees, clients, subcontractors, suppliers, and the general public. The published mission/vision statements of these three organizations are provided in Table 10.3. Other organizations with exceptional safety performance such as Hoffman Construction and Turner Construction have web pages that are dedicated to communicating the organization's safety-related missions and values.

Stating the organization's commitment to safety is a driver for the zero-injury goal which is pervasive among leading construction organizations. The zero-injury philosophy is rooted in the belief that construction projects can be successfully completed without human suffering. Many construction projects have now been completed without the occurrence of a single OSHA recordable injury. As one can see from Table 10.3, Skanska AB and J.B. Henderson Construction Co. specifically state that they are committed to an injury-free workplace in their mission statements. While many firms are experiencing significant improvements in their injury rates, even high-performing organizations such as those highlighted in this chapter have room for improvement. Hinze and Wilson (2000) showed that continuous improvement strategies implemented in an effort to meet zero-injury objectives have led to an additional reduction in injuries. The implication is that firms with

Table 10.3 Mission statements of highly-effective construction firms

McCarthy Building Companies, Inc. (www.mccarthy.com/about/mission-and-philosophy) EMR < 0.70	"At the end of the day, it's our people who make the true difference. Each person's commitment to excellence. Willingness to work in a team environment. Desire to do whatever it takes. How do we achieve this commitment from our owner-employees? By emphasizing and providing the following: • Tools, training and opportunities for growth • An environment that encourages and recognizes innovation • Challenging career opportunities • Exceptional financial rewards • Sense of family • A zealous commitment to safety and quality."
Skanska AB (www.skanska.com/en/About-Skanska/Our-targets) EMR < 0.70	"In addition to the financial targets – and as means to reach them – we have also adopted qualitative targets. The qualitative targets are expressed in Skanska's five zeros vision: 1. Zero loss-making projects. Loss makers destroy profitability and customer relationships 2. Zero accidents, whereby the safety of our personnel as well as subcontractors, suppliers and general public is ensured at and around our projects 3. Zero environmental incidents, by which our projects should be executed in a manner that minimizes environmental impact 4. Zero ethical breaches, meaning that we take a zero tolerance approach to any form of bribery or corruption 5. Zero defects, with the double aim of improving the bottom line and increasing customer satisfaction"
J.B. Henderson Construction Co. (www.jbhenderson.com/choose/mission.php) EMR < 0.60	"VISION J. B. Henderson Construction Co. is committed to achieving the highest level of quality and performance, to being known as the supplier of choice, and to building the future of families through jobs, training and community involvement. MISSION J. B. Henderson Construction Co. is a General/Mechanical Contractor. We deliver the highest quality and best value to our customers by integrating teamwork, dedicated project management, and outstanding craftsmanship. We create a desirable workplace, an injury-free environment and good jobs through equal opportunity. CORE VALUES Safety, Quality, Honesty, and Integrity"

good safety records can still make improvements by integrating safety-specific practices throughout planning, management, and project execution. While it is widely believed that zero risk is unachievable, continuous improvement leads to stable work environments with minimum risks.

Resource allocation

Rhetoric is not sufficient to establish or maintain a culture of safety. Managers of construction organizations must demonstrate commitment to safety objectives by funding safety-related activities and allocating a significant portion of their own time to active participation in regular safety program activities such as orientation and training, toolbox talks, and inspections. Resources must also be committed to hire and support full-time safety personnel. Some firms will allot at least one full-time safety professional to a project for every 50 employees on the project site. Management makes a tremendous statement about safety when a project with 500 workers has ten safety managers assigned to it as opposed to only one or two. Financial resources can also be effectively utilized to support safety lunches, rewards, and the purchase of personal protective equipment.

A recent study conducted by the writers focused on quantifying the investments made in regular safety programs by top construction organizations. The findings indicate that, on average, these firms invest approximately 2.5 percent of the total bid price on the foundational elements of the safety program. While the distribution of these funds to specific safety program elements depends on the project type and industry sector, the average distribution of funds is included in Table 10.4. These values include both first costs and ongoing costs and account for time spent by employees at all levels. Time spent by upper managers in safety-related activities as a separate line item in Table 10.4.

Employee evaluations and recognition

It is common for supervisory personnel and lower-level managers to be evaluated on a regular basis. Favorable results of these evaluations are typically associated with promotions and salary increases. Therefore, the basis of these evaluations is a very important issue. If

Table 10.4 Allocation of safety budget to specific injury prevention strategies

Safety program element	Percentage of total safety investment
Subcontractor management	4%
Upper-management support	6%
Employee involvement	8%
Job hazard analyses (JHA)	7%
Training and regular meetings	11%
Inspections	3%
Safety manager	6%
Substance abuse programs	19%
Written plan	7%
Committees	6%
Orientation and training	13%
Recordkeeping	4%
Emergency response plan	7%

profitability is the sole criterion, it is understandable that the supervisor being evaluated will have primary efforts focused on profitability. On the other hand, if profitability is a primary factor, conditioned on the requirement that safety must be maintained, the message to the supervisor of what is important is quite different. It is imperative that management make it clear that safety remains a core value and that this will be considered when evaluations will take place.

A controversial management technique related to evaluations is the use of safety incentives and disincentives. In a study conducted by Hinze (2002), 80 percent of construction firms had safety incentives and 86 percent had some form of negative reinforcement for unsafe work behavior. While incentives often promote safe work behavior, rewarding safety outcomes has been shown to encourage underreporting of injuries. Thus, if used, incentive programs should focus on rewarding leading indicators of safety performance such as management's observation of safe work behavior.

Studies have found, however, that sanctions for unsafe work behavior are effective for curbing such behavior. While verbal reprimands are largely ineffective, penalizing employees for unsafe work behavior in merit reviews and promotion decisions has been found to be extremely effective. Likewise, incentive programs can be effective in reducing the frequency of workplace injuries if appropriately structured. When structuring an incentive program, managers should focus on safe work behavior rather than outcomes (e.g. the occurrence of injuries). By focusing on the positive aspects of working safely, the incentive programs will promote safe behavior and discourage underreporting of injuries which can lead to systemic problems in the safety program.

Safety integration

In addition to upper-management support and employee evaluations, a critical element of a highly effective safety culture is the integration of safety within the various functions of the organization and in all phases of project delivery. Traditionally, safety management has been compliance-based and reactionary. Though some organizations have matured beyond this approach, there are many construction organizations that still follow this traditional paradigm. In order to meet the zero-injury objectives set forth by leading construction organizations, safety must be completely integrated.

Safety must be integrated throughout the various functions of an organization. Upper-level managers must recognize the difference between safety being housed within a particular department or division and safety being an integral component of every aspect of the company. If safety is viewed as belonging to a particular department or division, there is sense that safety is owned by that department or division. Instead, in the area of safety, every individual and every department must be responsible. There is no party involved in the construction of any facility that can truly say that safety is not their responsibility. Even if there is a safety department with a number of safety personnel, they are not the individuals ultimately responsible for jobsite safety. The safety personnel are to be viewed as resources that are to assist others in the performance of their responsibilities.

In addition to integrating safety within the organization, integration should also take place throughout the various phases of project delivery. Integration during the feasibility analyses and conceptual design can be achieved by involving knowledgeable safety consultants early in the project lifecycle. Owner organizations and designers should strive to avoid project characteristics that may pose exceptional or unique safety challenges. This awareness should be expressed during the selection of a project site, the development of the site layout, and the timeline for the project. For example, sites might be located where significant distances must be traveled to transport injured workers to medical facilities, the project might present potential toxic exposures, or the site might be suspected of harboring dangerous environmental conditions. These types of conditions should be addressed early in project planning. While it is not reasonable to make conceptual decisions based solely upon construction safety considerations, safety should be considered along with cost, schedule, and quality when performing multi-criteria decision analyses during feasibility studies.

A topic that has sparked particular interest in the construction industry during the last decade is the explicit consideration of construction safety and health during the design of a permanent facility. Termed "Design for Safety" (DfS), the technique has shown great promise in the reduction of occupational injuries and illnesses. While there are many barriers to effective integration of safety into the design phase, the technique has been shown to be viable (Gambatese *et al.* 2005). Commonly cited barriers include lack of safety knowledge on the part of designers, fear of increased liability for designers, industry fragmentation, and lack of incentives for designers. Gambatese (1998) demonstrated that, according to case precedent, consideration of safety in design does not increase a designer's liability. Further, to facilitate the design for safety movement, Gambatese *et al.* (1997) developed a 'DfS' toolbox with over 400 suggestions for designers.

DfS has progressed considerably further in the UK than in the US as a result of the 2007 Construction (and Design) Regulations (CDM). These regulations transfer some liability of construction worker safety and health to designers by requiring designers to "design-out" safety hazards. It is unlikely that similar legislation will be passed in the near future in the US. For the US construction industry to accept significant change regarding DfS, owners must first compensate designers for the additional time required to consider safety during their designs and take a more active role in promoting construction safety.

Some project delivery methods such as design-build (DB) and construction management at risk offer potential for enhanced input from constructors during the design phase. A common barrier to DfS in design-bid-build (DBB) projects is that the contractor generally cannot provide significant input until the contract documents have been developed. In DB and CM/GC projects, input can be obtained from the contractor during the constructability reviews that generally occur in the design phase. Furthermore, designers and contractors can collaborate to create a cost-effective, high quality facility that can be safely built in a timely manner. Coincidentally, it has been shown that both schedule and safety performance is better for DB than DBB projects.

Planning for safety

Safety must be considered early in the life of a project to ensure that it is effectively integrated. One of these early steps in a project is planning, which should not be undertaken without a specific regard for safety. Many decisions made during the planning stage of a project will have safety implications. This begins with the design of the site layout. Decisions of where construction roads will be routed, materials will be stored on site, workers will be permitted to park their vehicles, temporary utilities will be routed, tower cranes and other large cranes will be located, and how access will be gained to the site will directly impact the safety performance of a project. No planning should be done without the presence of safety personnel or without someone who can fully appreciate the safety implications of various planning decisions. This planning should also consider the implications of crowding on the jobsite, which can directly impact productivity and safety.

Rowlinson (2000) argued that integration of safety into planning starts with considering safety during the scheduling of a construction project. Scheduling for safety starts with a work breakdown structure and a logic diagram of activity precedence and involves the subsequent attachment of safety information to a critical path method (CPM) schedule or bar chart. Coble and Elliot (2000) and Kartam (1997) describe several types of safety information that can be attached to safety schedules including risk mitigation strategies, relevant training modules, and job hazard analyses. While the strategy of integrating safety data into project schedules does not reduce safety risk in its own right, the technique is effective for organizing safety knowledge so that it may be obtained quickly on an "on demand" basis.

Another scheduling strategy involves safety loading quantitative risk data into project schedules in order to identify periods of peak risk. In the same way that schedulers resource load a project schedule, safety data can be included in a work breakdown structure and subsequently plotted over time once the CPM schedule has been created. This technique can be used for multiple applications and is especially effective when safety managers are responsible for multiple projects. Specifically, schedulers can use risk data and safety knowledge to:

- avoid conflicting activities;
- overlay concurrent projects, analyze the risk profiles, and deploy safety managers to the highest risk project with the salient safety knowledge for the high risk activities;
- use float to level risk in an effort to increase predictability and avoid concurrent high risk activities; and
- use float to concentrate risk so that safety management activities can be strategically conducted immediately preceding and during high risk periods.

Integration of safety data into project scheduling and using relevant safety data to strategically sequence a project is a relatively new technique with great promise. This technique allows project managers to use available safety information to enhance decision making and, along with the DfS technique, is one of the best known methods for proactively managing safety.

Ethics

It has already been stressed that the promotion of worker safety makes good business sense. In addition, the promotion of worker well-being is a moral or ethical obligation of construction leaders. While there have been instances in which worker deaths have been labeled as being criminally-negligent homicides, the following comments will be confined to the ethical obligations of construction leaders to provide for the safety and well-being of the construction workers. If a supervisor realizes that a task can be performed in two ways, with one being clearly more hazardous, there should be no doubt in the mind of the supervisor of which method is to be employed. Worker well-being should always be a primary consideration. It is the right thing to do. The same thought process should be followed by estimators, schedulers, project managers, and others when two different alternative approaches to performing a task are envisioned. Shared ethical standards underpin a culture of safety. When safety is a core value, ethical decisions will become second nature.

Operations and execution

A project that has been well planned should progress smoothly. However, it must also be recognized that plans are often based on assumptions. In many instances, these assumptions may be accurate, but conditions or circumstances on construction projects are known to change. It is important to accommodate these changes. One way to do this is to be flexible at all times, with the realization that plans often must change to properly address the changing circumstances. Constant vigilance is required to continuously assess the planned work and the work actually being performed. When a change in the plan is warranted, there should be no delay in doing so in an effective manner.

In order for a project to be completed safely, effective communication is essential. Communication entails the transfer of information of various types. Early in the project, an assessment should be made of the major risks posed by the construction project. These risks are generally in the form of hazards that must be addressed in some manner. These hazards will include conditions such as excavation and trenching hazards, traffic hazards at the project entrance/exit, overhead power lines, the movement of large pieces of equipment, and so on. It is imperative that plans be developed to address these hazards. These plans must then be communicated to all individuals that might be exposed to those risks, including the public in some instances.

Despite extensive efforts by a firm to identify all major hazards prior to the start of construction, not all hazardous conditions can be envisioned before construction begins. In most projects, hazards do not exist until construction work begins. Thus, assumptions must often be made about the methods to be employed to perform certain tasks. For example, small excavations might be dug by machines or they might be dug by hand. Some of these decisions will not be made until construction work has begun. In fact, sometimes changes might be made in the work plan to accommodate changes in jobsite conditions. Small excavation work may have been planned to be performed with machinery, but excessive rainfall might cause a manager to change the plan to digging the excavation by hand to avoid making a muddy mess with the machinery. Such changes need to be communicated promptly to all individuals affected by such decisions. In some cases the changes will be minor, but if the excavation is deep and workers will be required to work in the excavation, then more serious considerations are involved.

Whenever a new risk is identified, a corresponding safe work plan be devised. The first objective of the work plan will be to eliminate the hazard or to rearrange the work so that workers are not in harm's way. The last resort would be to equip the workers with the proper safety devices so that the hazard is controlled.

Although a project might be undertaken with the expectation of incurring no injuries, it is important to be prepared for an unlikely accident in which a worker might be seriously injured. When an injury occurs, the objective is to provide prompt treatment of the injured individual. To improve the effectiveness of any response to an injury it is important to have a clear understanding of the roles of the various individuals that might be involved in obtaining treatment for the injured individual. Thus, if an injury does occur an immediate network is activated in which a supervisor or safety manager takes charge of the situation by first assessing the nature of the injury and the appropriate response to the injury itself, and a vehicle is acquired for transportation (whether it be a company vehicle or an ambulance for very serious incidents). Since a company with a zero-injury philosophy will probably encounter few serious injuries, the response to injuries may not be a regular occurrence. To ensure the efficient response to serious injuries, the proper sequence of steps to be followed should be reviewed periodically and regular rehearsals might be warranted.

After transportation has been acquired for an injured worker, attention should then be given to addressing the cause of the incident. This may be performed by a safety manager, ideally with the assistance of the immediate supervisor of the injured worker, and will involve an investigation of the circumstances that resulted in the incident. This effort will include a thorough evaluation and documentation (digital photos, etc.) of the work area and interviews with witnesses. It is important that witnesses be interviewed as soon as possible, as memories can fade quickly and the facts can quickly become distorted if workers discuss an incident between themselves. The end result of this investigation should be a report in which steps are outlined by which similar future incidents will be avoided.

Role and responsibilities of upper-level managers

A central theme of this chapter is that managers of construction organizations play a pivotal role in the development and maintenance of a strong safety culture. In fact, the demonstrated commitment of upper-level management can be well exhibited through their active participation in regular safety activities and by allocating sufficient resources to support those programs designed to reduce injuries. It must be recognized that not all participation activities by upper management are effective. The participation must be of a nature that truly communicates a deep commitment to safety. It is crucial for the leaders of construction organizations to understand their role in safety management activities.

- Managers should regularly participate (in person) in safety inspections of active worksites, toolbox talks, training and orientations sessions, safety committees, and preconstruction safety planning. Active participation in such management activities shows employees at all levels that the organization takes safety very seriously. As a corollary, upper-level managers who do not participate in safety-related activities but show an obvious concern for productivity, quality, and cost send the message to their employees that safety is relatively unimportant.
- Upper-level managers should participate in regular safety efforts and treat others in the discussions as equal partners in safety. This means that managers must participate by listening to the ideas of all individuals and share ideas only when they add significant value to the discussion. Upper-level managers should not dominate these meetings. Rather, they should show an interest in the ideas, opinions, and experiences of their employees, i.e. oftentimes it is the worker with many years of experience who may have the best ideas on safety. Further, managers should utilize their experiences to develop and fund innovative safety efforts. If a safety manager is employed in a firm, this individual should lead safety management activities while project managers and upper-level managers should participate as members of the crew.
- Managers should involve workers at all levels in safety planning and management activities. No individual should be left out of safety planning . This means that workers of all levels are not only trained, observed, or lectured to, but they are to actively participate as inspectors, committee members, and co-leaders of safety meetings.
- Managers must observe active worksites and personally acknowledge safe work behavior. Further, managers must hold workers accountable when they elect to work in an unsafe manner. Most importantly, however, managers must serve as pristine role models of safe work behavior and hold one another accountable. Workers should know that no one is

above safe work behavior and should hold even the highest-level managers accountable for their behavior.

- Managers should be evaluated according to the safe work practices of their employee base. Evaluations should focus on observable performance, not only on injury rates and statistics, as injuries occur relatively infrequently. Laborers, foremen, superintendents, and project managers should also be evaluated on the basis of their work behavior and the behavior of their subordinates. Safety should be a criterion for all promotions, raises, and company awards.
- Managers should inform workers that they have unilateral power to stop work if an uncontrolled hazardous exposure exists. This worker empowerment must not be questioned, but it must always be supported and reinforced. This should be a risk-free power given to all employees. Accordingly, there must be a formal method for communicating a hazard that warrants a stop work order, a procedure for investigating and controlling hazardous exposures, and subsequent countermeasures to prevent the reoccurrence of such an exposure in the future. Employees who identify and communicate such hazards should be recognized publically by management for their insights, awareness, and effort.
- Managers should conduct frequent safety audits using observational techniques, safety climate surveys, and evaluating all accident reports. Managers should also pay close attention to high frequency, low severity injuries such as strains, first aid injuries, workers who have persistent discomfort and pain as a result of their work, and close calls. Frequent low severity injuries are generally a sign of systematic safety problems that have the potential to result in a high severity injury or catastrophe, i.e. they are clear signs that there is a weakness in the safety process.
- A relatively difficult but highly effective safety management technique is to educate clients, subcontractors, and suppliers about the importance of safety and the expected work behavior on site. Because of the complex nature of construction projects there may be many independent firms working on a project at any given time. As a result, the work practices of all individuals contribute to the level of site safety. Site safety is difficult to manage due to contractual language, liability and industry fragmentation. Nevertheless, setting clear expectations for all individuals who plan to work on or visit the worksite is essential to the development of a strong safety climate and accident prevention. It is important to note that many of the aforementioned safety strategies will have little impact if employees are forced to work in an environment dominated by employees of other organizations who have poor values and beliefs with regard to safety.

Advanced management strategies

Effective managers stay abreast of emerging innovative products, processes, technologies, and services. It is the role of managers to "push the envelope" with respect to safety to gain or maintain a competitive advantage. Several emerging safety management strategies are identified and briefly described below. This is by no means a comprehensive list of emerging construction safety strategies; rather, these are a few examples of emerging strategies that show promise.

Safety knowledge management as a strategy for continuous improvement

Though knowledge management applies to all aspects of managing construction projects, the management of safety knowledge is particularly important. Identifying safe work practices for construction tasks and effective strategies to reduce hazardous exposure is at the essence of effective safety knowledge management. Information that can contribute to such knowledge is available from many sources that are external to any one organization. For example, the Occupational Safety and Health Administration (OSHA) offers employers a wealth of high quality safety information free of charge. If invited, OSHA representatives or consultants will conduct consequence-free inspections or training. Additionally, there are societies dedicated to the creation and dissemination of safety knowledge such as the American Society of Safety Engineers. Decades of safety knowledge is also available in books, journal publications, and trade journals. Safety consultants also represent a source of safety knowledge that can be quite extensive.

Internal management of safety knowledge is more complex. In order to effectively transfer knowledge among individuals, work crews, projects, regions, and countries, formal knowledge management systems must be established. Some individuals refer to such a system as a lessons learned program. First, tacit and explicit knowledge generated by employees must be captured with formal, consistent processes. An example of an effective knowledge capturing system is a Job Hazard Analysis (JHA). Generally, JHAs include a description of tasks associated with a specific construction process, the risks associated with the tasks, and effective strategies for accident prevention. While some organizations use JHAs prepared by consulting organizations, most utilize the collective experience of their employees to create the content for their JHAs. The creation and continuous improvement of JHAs offers an opportunity to leverage the wealth of safety knowledge gained or acquired by workers and crew supervisors. Other examples of opportunities for capturing safety knowledge include the development of project-specific safety plans, safety inspections, and accident analyses. All of these processes offer opportunities to involve workers of all levels in safety management and planning which is likely to result in increased buy-in and compliance with safe work practices.

Internal and external safety knowledge must be consistently stored in strategic locations. Possible storage locations include online databases, written safety plans and procedures, JHAs, accident reports, case studies, relevant journal articles that describe safe work behavior, diagrams, safety committee reports, and descriptions of safety innovations. In the digital age, storage of a large volume of safety information is possible. However, to be effective, this knowledge must be sorted appropriately to ensure easy access to relevant, high-quality information. The information must be easy to access, searchable, and quickly return relevant high-quality information. If the knowledge storage system is cumbersome, slow, or confusing, workers or supervisors are much less likely to use the system.

Finally, knowledge must be transferred to and easily accessed by all employees of the organization to be effective. Many high-performing organizations have an intranet that includes safety information that is searchable by keyword. Knowledge can be diffused throughout the organization by involving all individuals in safety management activities such as the development of new safety plans, safety constructability meetings, safety committees, orientation and training seminars, safety inspections, and toolbox talks. Essential safety knowledge can also be disseminated throughout the organization through newsletters that accompany regular paychecks, presentations made at receptions, company-wide addresses from upper management, signage, and other methods.

When used appropriately, knowledge management systems can improve business performance (Carillo and Chinowsky 2006). Safety knowledge management is an essential component of an organization's overall knowledge system that is typically overlooked. Many of the high performing organizations seek to formalize their safety knowledge system to enhance their safety culture. Some organizations go as far as to sponsor academic research, participate on local, national, and international construction safety committees and communities of practice. These organizations utilize these opportunities to stay at the cutting edge of the industry.

Integration of safety into Building Information Models (BIM)

Building Information Modeling (BIM) is a strategy that involves attaching quantity, cost, schedule, progress, spatial, material attributes, and other forms of data to physical building elements in three dimensional computer models. BIM is an emerging technique that is particularly useful for estimating, scheduling, and project controls for complex projects. Researchers are currently establishing safety data for inclusion in BIM. Such data will include quantitative safety risk data, conflicts among building elements and worker activities, and conflicts between physical elements and work processes. In the future, it is expected that BIM models will be used for safety management by providing workers with real-time information by which to make changes in the hazardous conditions that exist on the current state of the project.

Real-time proximity alert systems

Another emerging strategy is the use of radio frequency identification (RFID), ultra-wideband (UWB), laser scanning, and other sensing technologies to provide workers with real-time proximity warnings. This technology shows particular promise in the highway sector where there is a relatively high rate of serious injuries from contact between mobile heavy equipment and workers on foot. RFID technology, for example, provides the workers on foot and the equipment operators with audible warnings when a given threshold distance has been reached. While UWB and laser scanning are more expensive technologies, they produce high quality data and can be used to track workers through a work zone. Much development and pilot testing is required before these technologies will see widespread use.

Visualization and simulation

Finally, as construction simulation programs such as STROBOSCOPE continue to mature and diversify, these technologies can be used, along with actual safety data, to simulate construction conditions. These simulations could be used to illustrate safety hazards expected on complex, high risk work sites that can be observed by workers and supervisors in a risk-free environment, thus safely preparing workers and managers for future hazardous exposures. Simulations could also be used for training purposes for new hires and for workers who transfer to a new site. Proactive safety planning has been shown to be one of the more effective management strategies, as this alerts others to the expected hazards before they actually encounter the hazards.

Concluding remarks

Jobsite safety has evolved incredibly since the days when a safety program consisted exclusively of holding a toolbox safety meeting once a month. The evolvement of construction safety has been steady and it has been effective. The OSHA recordable injury rate in the mid-1970s was about 20 recordable injuries per 200,000 worker hours. This rate has now dropped to less than 6.0. The fatality rate has also declined in a dramatic manner.

Many aggressive and proactive companies who have embraced the zero-injury philosophy have reduced their injury rates to less than 1.0, with some projects being completed without sustaining any recordable injuries. This does not mean that they have come to grips with the safety problem and fully solved it. The safety challenges will continue, even for those firms with world class safety records, as they will continue to hire workers who have worked for firms where injuries are an expected "part of the job" and they will continue to hire subcontractors that have become established in an environment where safety is not stressed in an aggressive or proactive manner. In short, the safety challenges will continue for all firms. Over time, the entire construction industry will continue to improve its safety performance. It is only through the diligent efforts of committed top managers that these efforts will be successful.

As long as workers are exposed to hazards on construction sites, the safety efforts of companies must be sustained. The trend of the past few decades has shown good promise for continued improvements in safety performance. At the core of any company's safety efforts must be the realization that the most valuable resource on the construction sites is the workforce. That workforce warrants preservation and protection. All construction workers deserve to work in an environment where they can work productively and return safely to their homes at the end of every working day. A manager who does not subscribe to this fundamental premise has failed.

Key strategies

Strategic safety management begins with understanding the impacts that occupational injuries have on business performance and recognizing the most cost-effective injury prevention activities. However, world-class safety performance requires that organizations establish a culture of safety where employees share the attitude that safety is paramount and that safe work behavior is expected of everyone at all times. Achieving a culture of safety requires demonstrated commitment of upper managers, allocation of adequate resources to achieve clearly-defined and achievable goals, integration of safety with all other project management activities, and inclusion of safety leading indicators as component of employee evaluations. Other essential strategies include instituting a safety knowledge management system that captures, stores, and disseminates key ideas and exploiting opportunities to leverage emerging technologies to prevent injuries.

References

BLS (Bureau of Labor Statistics) (2009), *Census of Fatal Occupational Injuries Summary, 2007*, Economic News Release, online, available at: www.bls.gov/news.release/cfoi.nr0.htm [accessed August 5, 2009].

CPWR (Center for Construction Research and Training) (2008), *The Construction Chart Book*, CPWR: Silver Springs, MD.

Carillo, P. and Chinowsky, P.S. (2006), "Exploiting knowledge management: The engineering and construction perspective," *Journal of Management in Engineering*, 22(1): 2–10.

Coble, R.J. and Elliot, B.R. (2000) "Scheduling for construction safety," in R.J. Coble, J. Hinze, and T.C. Haupt (eds.), *Construction Safety and Health Management*, Prentice Hall: Upper Saddle River, NJ, pp. 43-57.

Coble, R.J., Haupt, T.C., and Hinze, J. (2000), *The Management of Construction Safety and Health*, A.A. Balkema: Rotterdam.

Everett, J. and Frank, P. (1996), "Costs of accidents and injuries due to the construction industry," *Journal of Construction, Engineering and Management*, 122(2): 158–164.

Fredricks, T., Abudayyeh, O., Choi, S., Wiersma, M., and Charles, M. (2005), "Occupational injuries and fatalities in the roofing contracting industry," *Journal of Construction, Engineering and Management*, 131(11): 1233–1240.

Gambatese, J.A. (1998), "Liability in designing for construction worker safety," *Journal of Architectural Engineering*, 4(3): 107–112.

——, Behm, M. and Hinze, J.W. (2005), "Viability for Designing for Construction Worker Safety," *Journal of Construction, Engineering and Management*, 131(9): 1029–1036.

——, Hinze, J.W., and Haas, C. (1997), "Tool to Design for Construction Worker Safety," *Journal of Architectural Engineering*, 3(1): 32–41.

Gibb, A.G., Tubb, D. and Thompson, T. (1995), *Total Project Management of Construction Safety, Health and Environment*, second edition, Thomas Telford: London.

Hallowell, M.A. and Gambatese, J. (2009) "Construction safety risk mitigation," *Journal of Construction, Engineering and Management*, 135(12): 1316–1323.

Hill, D.C. (2004), *Construction Safety Management Planning*, American Society of Safety Engineers, Des Plaines: IA.

Hinze, J. (2000), "Incurring the costs of injuries versus investing in safety," in R. Coble, R. Hinze, and T. Haupt (eds.), *Construction Safety and Health Management*, Prentice Hall: Columbus, OH, pp. 23–41.

—— (2002), "Safety incentives: Do they reduce injuries?" *Practice Periodical on Structural Design and Construction*, 7(1): 81–85.

—— (2006), *Construction Safety*, Prentice Hall: Englewood Cliffs, NJ.

—— and Pannullo, J. (1978), "Safety: function of job control," *Journal of the Construction Division*, 104(2): 241–249.

—— and Wilson, G. (2000), "Moving towards a zero injury objective," *Journal of Construction Engineering and Management*, 126 (5): 399–403.

Jaselskis, E., Anderson, S.A., and Russell, J. (1996), "Strategies for achieving excellence in construction safety performance," *Journal of Construction, Engineering and Management*, 122(1): 61–70.

Kartam, N.A. (1997), "Integrating safety and health performance into construction CPM," *Journal of Construction, Engineering and Management*, 123(2): 121–126.

Liska, R.W. (1993), *Zero Accident Techniques*, The Construction Industry Institute, The University of Texas: Austin, TX.

Mearns, K., Whitaker, S. M., and Flin, R. (2003), "Safety climate, safety management practice and safety performance in offshore environments," *Safety Science*, 41(1): 641–680.

NSC (National Safety Council) (2006), *Accident Facts*, NSC: Itasca, IL.

Neal, A., Griffin, M., and Hart, P. (2000), "The impact of organizational climate on safety climate and individual behavior," *Safety Science*, 34(1–3): 99–109.

Rajendran, S. (2006), *Sustainable construction safety and health rating system*, Ph.D. dissertation, Oregon State University: Corvallis, OR.

Rowlinson, S. (2000), "Human factors in construction safety: Management issues," in R. Coble, J. Hinze, and T. Haupt (eds.), *Construction Safety and Health Management*, Prentice Hall: Columbus, OH, pp. 59–85.

Sawacha, E., Naoum, S., and Fong, D. (1999), "Factors affecting safety performance on construction sites." *International Journal of Project Management*, 17(5): 309–315.

Tarrants, W.E. (1980), *The Measurement of Safety Performance*, Garland STPM Press: New York.

11 Corporate social responsibility

A strategic and systematic solution to global poverty

Stephen Darr and Anthony D. Songer

Introduction

stra·te·gic (strə-tē'jĭk) adj.
    a.   Important or essential in relation to a plan of action
    b.   Highly important to an intended objective
               The American Heritage® Dictionary of the English Language, fourth edition,
                          Copyright © 2009 by Houghton Mifflin Company.

sys·tem·at·ic [sis-tuh-mat-ik] adj.
    a.   Having, showing, or involving a system, method or plan
    b.   Given to or using a system or method; methodical
    c.   Arranged in or comprising an ordered system
                    Random House Dictionary, © Random House, Inc. 2009.

Nestlé does it. PricewaterhouseCoopers does it. Accenture does it. If these global giants can do it, why shouldn't we?

The *Harvard Business Review* featured in December 2006 an article titled "Strategy and Society: The Link Between Competitive Advantage and Corporate Social Responsibility" (Porter and Kramer). One of the examples was Nestlé. Once criticized for promoting formula over breast milk in developing countries, today Nestlé is making a difference and making a profit. Responsible social practices along with production goals led Nestlé to establish a dairy operation in one of the poorest rural areas of India. Nestlé's objective was not targeting poverty per se but, more strategically in this instance, raising the standard of living on rural farms to a level that helped the firm increase local dairy production "50-fold" while at the same time providing huge economic benefits to the region How did they do that? They introduced experts in a variety of fields to support not just the core business product, but also the people who served as Nestlé's producers.

You might not expect to find a major global accounting firm in the poorer neighborhoods of Belize City, but PricewaterhouseCoopers (PwC) is encouraging prospective new employees to see part of their job as "giving back" and thus involves staff and interns with young people in an innovative partnership with the Peacework Development Fund, Inc., (www.peacework. org) around educational opportunity, youth leadership, and economic development. To PwC, corporate social responsibility means being strategically aware of the broader economic development needs of communities in which they have a presence, and theirs is a big presence. Their program called Global Communities adds a humanitarian element to the PricewaterhouseCoopers operation in 151 countries and becomes part of a wide range of services around economic development, accountancy, and general business needs. Affiliated offices all over the world are involved in community projects. For instance, the

> PwC South Africa office decided to improve the lives of rural women by establishing the Business Skills for South Africa (BSSA) foundation. The initiative provides business and entrepreneurial skills to disadvantaged communities, with the specific aim of empowering women in isolated rural communities. They are given the training to generate their own income and ultimately make a meaningful contribution to South Africa's economy. This unique approach to a community issue has been recognised by the South African Black Management Forum for skills development and capacity building.
>
> (www.pwc.com/gx/en/corporate-responsibility/global-communities-recognition-awards.jhtml)

Virtexco is doing it. Yes, Virtexco of Norfolk, Virginia.

Bob Wells, co-owner of Virtexco, wanted to give back, not only to his alma mater, Virginia Tech, but also make a difference in the world. By getting the university's building construction department involved in a global service project, Bob was able to support building construction students in the act of service in their chosen profession, inspire other companies on the department's industry board, and perhaps even find a few stellar graduates for his construction firm. In March 2008, Bob took faculty and students, twenty-three of them as a matter of fact, from the Myers-Lawson School of Construction to the rural farm labor community of Bella Vista, Belize. They made extensive improvements to Our Lady of Bella Vista Catholic School. In a village of 7,000 primarily immigrant farm worker family members, the school serves 1,084 children. Bob's generosity paid off a thousand-fold.

If Virtexco can do it, we can! Make a difference in the world, that is.

Corporate social responsibility is in vogue, but as a general concept it's far from novel. The idea of making money and being a good citizen at the same time only makes sense. Most businesses are built on relationships. These relationships are not only with individual consumers but with people in other businesses, industries, charities, and public service organizations. With apologies to John Donne, "no business is an island entire to itself." While the idea of developing a truly strategic, business-savvy approach to good citizenship may be attractive to corporate executives, it is difficult to commit hard-earned cash to social projects when it is not at the core of their business. Times are bad enough; why invest in activities with intangible or distant financial benefits? This is true in all businesses but more so, we suspect, in the task-oriented fields of engineering and construction.

Global poverty and the industry potential

If an interstellar traveler could glide through billions of light years of space while passing millions of spectacular supernovas and clusters of spiraling galaxies, each with hundreds of millions of their own stars and planets, to miraculously arrive at this tiny blue and green rock hidden in the Milky Way, called Earth, that traveler would find an almost infinitesimally tiny sphere with a radius of only 3,949 miles inhabited by nearly seven billion human beings. While many of those human beings have fairly comfortable living conditions and access to shelter, food, and clean water, another one-third of those people on that very tiny sphere live on less than $2 per day. Amazingly, half of *that* population survives on less than $1 per day. One in every five children born in that population will die before they complete the first year of life. Those who survive that first year are malnourished and suffer from disparities in nutrition, health care, education, housing, and human rights. Our interstellar traveler might wonder why there are such wide disparities in living conditions among the occupants of this planet.

While exploring this tiny orbiting rock, our space traveler might also come across some large infrastructure projects run by the French company VINCI or an American firm named Bechtel. Bechtel and VINCI are two of the world's largest construction contractors. The annual revenues of VINCI alone exceed the gross domestic product of 112 of the world's 179 major countries (www.vinci.com). Connecting the two, a powerful corporation with the vulnerable poor, may be one of our most urgent and timely challenges in the next twenty years. Doing so, could change the face of poverty by lifting millions from their captive states of destitution, hopelessness, and violence. Applied strategically and systematically, their resources would inevitably enhance the human spirit, expand financial capacity, foster entrepreneurship, build infrastructure, and increase productivity among tens of millions of people, just as we are observing today in commerce in the People's Republic of China and the IT industry in India.

Engineering and construction's contribution potential

In *The End of Poverty* (2005), Columbia University's Jeffrey Sachs suggests two main sectors in the solution to global poverty – *investment and infrastructure* (p. 232).

To an accountant, a contractor, or any other professional, this prospect is both a formidable and amazing opportunity.

Just think: the apprentice engineer has within her professional prospects one of the two secrets to ending global poverty. But does she have the experience and drive to help solve a worldwide dilemma? She does *if she's been there* and listened to people talk about their needs, and understands from personal observable experience the dynamics that perpetuate poverty. Only then can she begin to piece together the solution to one of the most insidious of human conditions. "Giving back" by engineers and contractors is the infrastructure aspect of the solution.

Investment

By investment, Sachs wants to "introduce new capital" in the form of human, institutional, and financial resources where poverty is widespread, thereby prompting dramatic new human, business, social, and public action in those villages. The most significant capital is human – individuals who are part of corporations, universities, and other institutions collaborating across disciplines like education, health care, and business. People bring investment, be it corporate or otherwise and, as Sachs emphasizes, the missing piece in development in these villages is often simply a modest amount of funding. A small investment from outside the village typically makes an enormous difference where a little means a lot.

Moderate outside investment is critical. Remember the adage, "the rich get richer and the poor get poorer?" That is exactly what is happening in the global village. It is virtually impossible for the poor alone to overcome the combination of economic obstacles standing in their way. It may require only a very modest investment to start or sustain a small enterprise, but even a tiny investment is simply too great an amount for the poor to afford because it is more urgent to use those meager resources for food and shelter.

In developing communities, it is vital to have even modest participation and funding from outside sources. Even limited involvement by others can be a huge boost to a local village. It can launch a project, get a project "over the hump," or put the finishing touches on a venture. It is potentially better to have a small investment so that a windfall does not overwhelm the community. A modest investment is more appropriate to the typical scale of development, more consistent with surrounding needs, less prone to creating conflicts and conflict of interest, and far more manageable. It is more helpful having *people committed to a village partnership* over a *sustained* period of time at levels of involvement and funding that *match* each development objective rather than sending millions in World Bank support. Face-to-face commitment, continuity, and collaboration are more important than money.

Introducing an unpretentious amount of financial support or collaborating with leaders from inside the village on improving the water system, electrification, school construction, farm irrigation, clinic renovations, or bridge construction simply means *going to these communities* and participating in the daily life of the people, understanding the culture, and becoming more intimately familiar with the real local needs, capacity, and resources. Regardless of the discipline or approach, being personally and intimately involved is vital in the business of infrastructure.

Half of "giving back" is "being there." Being in the village is more than symbolic. Collaboration produces ideas, multiplies resources, and motivates people to move beyond their current situations. Jerry Aaker, formerly of Lutheran World Relief and Heifer Project International, described this phenomenon in *Partners with the Poor*. He talked about working "alongside" people of developing communities, not simply bringing money or "helping" but rather accompanying people through their own process of self-development.

Throughout twenty years of working with community and village partners all over the world, Peacework has banned the word *help* from its organizational vocabulary. True to Aaker's theory of "accompaniment," we simply work alongside those who already have insights and knowledge of development objectives in their own community. This work is not a matter of helping, it is a matter of walking and working alongside those who are poor and marginalized in our world on their journey to self-reliance and toward their capacity for self-development. That is really all we are able to do as guests in someone else's home and community. It seems simplistic, but in fact everything boils down to the concepts of neighbor and village.

The "accompaniment" concept is crucial for corporate social responsibility. Accompaniment leads to empowerment among those who do not have even the most basic material resources. Empowerment, then, leads to development initiatives that truly meet the needs of the local community and more than likely also result in profitable endeavors for the corporation. Some of those endeavors may be profitable only in terms of positive public relations, however; the corporation can also be strategically engaged in the local village so that it ultimately benefits both the village and the bottom line. That's what happened with Nestlé in the rural villages of India.

Roberto Noensie and Mervin Jabaraj, students of the Walton College of Business at the University of Arkansas, took the concept of micro-loans and applied them to educational needs rather than small businesses. The micro-loan concept itself was brilliant. In 2006, Muhammad Yunus won the Nobel Peace Prize for his widespread expansion of micro-financing for small businesses among the poor. The story goes that

Yunus started this innovative credit program in 1974 when he loaned "the equivalent of $27 out of his pocket to 42 women in the village of Jobra who had a small business making bamboo furniture." Two years later, he founded the Grameen Bank "pioneering a new category of banking ... The program has enabled millions of Bangladeshis, almost all women, to buy everything from cows to cell phones in order to start and run their own businesses ... and now, similar micro-credit projects have helped millions around the world lift themselves out of poverty." By 2007, the first $27 Yunus pulled from his pocket had resulted in $6.38 billion in loans to 7.4 million people, 96 percent of whom are women, and new ventures in agriculture and business development among the poor. The typical loan amount? $200. The result? Opportunities that will change lives.

(Stefan Lovgren, *National Geographic News*, October 13, 2006)

Peacework introduced the global accounting and business firm of PricewaterhouseCoopers to four schools in Belize City. Two of those schools are situated in the Queens Square and Port Loyola sections of the city where residents are struggling with poverty and increased gang activities. PricewaterhouseCoopers, in turn, asked their summer interns from business schools around the country to step up to the challenge. Within months, the firm's Office of Corporate Social Responsibility in Chicago and Recruiting Office in New York delivered over four tons of supplies to four schools, built and equipped four new school resource centers with current technology, excavated the school yards to prevent flooding, engaged 100 interns and forty staff in four sports and entrepreneurship camps with over 400 Belizean children, contributed over $11,000 from their own pockets to school costs, participated in a virtual exchange program via email between 420 children and over 500 of the firm's interns, and provided $35,000 in scholarships to 115 youth who may never have had the chance to advance beyond the equivalent of the sixth grade. After the scholarships were announced, one of the young recipients with perfect confidence and poise assured her audience that she will one day be Belize's first female prime minister. The typical scholarship? $250.

A micro-loan of $200 or a scholarship of $250 does not solve the whole world's problems, but it goes a long way in overcoming the almost insurmountable obstacles *one* young person faces at that stage of their lives and in that setting, and opening doors that might otherwise have remained shut. It is a powerful resource for people living in some of the most difficult conditions of poverty in our world when accompanied by a sustained commitment to alleviating poverty over the long haul. The trick is developing and maintaining the village's capacity to meet educational, health care, and other needs for years. This will be required to reverse long periods of disparity and inequity sustaining the corporation's financial capacity and managerial commitment to the village partnership.

Peacework started the Peacework Village Network in 2006. Like many innovations, the concept is simple. Many other development organizations work in their area of specialty such as medicine or housing. Yet the roots of poverty are complex and the contributing factors include policies and disciplines that corporations can address by joining forces with other business ventures, non-governmental organizations, or public service organizations – business, education, agriculture, health care, engineering, and law. Within these fields are hundreds of businesses, services, and engineering applications, each just as vital to overall economic development as the other. The thrust of the village network is to introduce all these resources simultaneously into the village while listening to and coming to understand the real-world insights and experience of local community leaders.

The presumption here is that the insights of the rural community and the resources of corporations applied across a variety of fields will greatly increase the capacity of the village for effective economic growth. That partnership will also broaden the horizons of every participant (interns, staff, management, executives, educators, chambers of commerce, and even corporate clients) whose innovations, ideas, and collaboration foster social change.

Infrastructure

By infrastructure, Sachs recognizes that the integrated function of communication technology, transportation, construction, sanitation, mechanical, and energy systems in the community are essential to development. These systems correspond to the essential parts of the anatomy that all have to work together for the organism to survive and thrive – skeletal, circulatory, musculoskeletal, nervous, respiratory, integumentary, and vestibular. Each system is interdependent on the others for function, protection, and development of the whole.

Engineers and builders, ostensibly, ensure that the community has adequate structural capacity to educate its children, feed its population, provide health care, move essentials from place to place, and communicate inside and outside the village. The systems are symbiotic. As the musculoskeletal system will not work independently from the circulatory or skeletal systems, schools will not fulfill their functions without institutions that ensure human rights and equitable application of laws, access to food and education about nutrition, rights to medical care, and meet the region's transportation and communication needs in one consolidated campaign against poverty.

True to the Village Network concept, the infrastructure must be designed *in response to* the needs, use, scope, and management capacity of the village. Therefore, the bridge over the Stann Creek in Dangriga, Belize is, of course, designed with input on the weight and volume of local traffic, the size of the local population, and available finances. Water systems are designed according to the scale of distribution and source. Medical facilities are designed according to the projected case load and the range of services the clinic will provide. Schools are designed for student capacity, available teaching staff, and services such as special education or a computer lab. All of these projects in the developing world are subject to severe funding constraints. Few design features are determined by the outside engineers but by the village and available resources.

Sachs referred to his work with the village of Sauri in Kenya. "The village could be rescued," he wrote, "but not by itself" (Sachs 2005). "Survival depends on addressing a series of specific challenges: nutrient-depleted soils, erratic rainfall, holoendemic malaria, pandemic HIV/AIDS, lack of adequate education opportunities, lack of access to safe drinking water and latrines, and the unmet need for basic transport, electricity, cooking fuels, and communications. All of these challenges can be met with *known, proven, reliable, and appropriate* technologies and interventions" (Sachs 2005). The key phrase here is that "the village can be rescued … but not by itself" and the hypothesis is that infrastructure along with investment is the apparent solution to poverty. The premise is that infrastructure and investment solutions will be much more effective if they are applied with varied resources from the corporation or corporations in partnership with universities.

The multi-disciplinary nature of the village partnership does not diminish the potential impact of separate projects. School gardens can be an integral part of the primary school curriculum as well as micro-loans for education and small enterprises. These alone can be powerful tools of development, yet the deliverables, the outcome, will be more sustainable and effective when supported by other disciplines. That is what Nestlé found out when they started expanding the dairy initiative in rural India. Local farms needed minimal incentive and resources to achieve the deliverables that Nestlé required and produced very profitable results. Yet, the expansion required not only managers but veterinarians and educators. The Peacework school garden is a great example.

At St. Matthew's Anglican School in Pomona, Belize, over 100 children attend the two infant grades (kindergarten) and six standard grades (primary school). One might think that the biggest problem would be a lack of pre-school preparation, inadequate support from parents, low funding, the unavailability of current computer technology, or problems with the facilities, but the biggest problem is that the children come to school hungry. Hunger is an enemy of physical growth, psychological development, socialization, and effective education.

Ironically, the school is situated in a fertile plain in the Stann Creek valley where citrus groves cover the terrain providing tremendous opportunity. These kids could be planning, planting, growing, harvesting, learning about plant physiology and nutrition, and even setting up a vegetable market with, yes, a simple school garden. School gardens are not new to this world, but they are new to Pomona.

The uniqueness of the idea is the involvement of a partner. University of Arkansas students and faculty in agriculture, engineering, business, education, and literacy could all work together with the kids, teachers, and parents on this amazing project. Agriculture students worked alongside the kids on planning, plowing, and planting. Engineering students constructed a water collection system using the school roof, a collection tank, valves, and pipes to irrigate the garden while fencing the perimeter. Education students and school teachers devised a curriculum for the garden program and worked with parents and children to institute a family literacy project around the garden. Business students worked with kids on a business plan, how to market the produce from the garden to generate funds for seeds, tools, and transportation. Two years into the program, it was apparent that the occasional contact between the university and the school was not sufficient to keep the cycle of planting and harvesting working well. In the third year, the program added funding for a garden manager. His oversight, a simple addition to the partnership equation, made all the difference. The cost is a very modest $25 a month.

This strategic use of a very small amount of money in a very small project gave kids the encouragement and teachers the support needed to save literally hundreds if not thousands of dollars in annual re-tilling, re-seeding, clearing, weeding, security, and maintenance not to mention the follow-through on the educational plan devised by the partnership. Before the garden was finished, the Minister of Agriculture asked if the same project could be undertaken for all of Belize schools. Imagine the kick start that this modest corporate support would give to this project. The project could feed a generation in this community. The model could feed a generation worldwide.

The corporation

The corporation's social responsibility considerations are numerous. The firm would be committing valuable financial and human resources outside of its core mission. The risk of working with the poor and with remote areas of the world is high for a business built on reputation, stability, and tightly run management. If we are to get around these obstacles, the corporation has to accept social responsibility as a genuine part of its core vision and purpose. How the corporation defines its commitment to social responsibility will reverberate throughout the firm's structure, personnel, and clientele.

Not only are there internal corporate pressures, society is increasingly expecting profit-making firms to include some form of "giving back" to the community. Porter and Kramer note that there are four "prevailing justifications" for corporate social responsibility (CSR) – "moral obligation, sustainability, license to operate, and reputation" (Porter and Kramer 2005). In the long-run, charitable enterprises that benefit society and increase the population's capacity to participate in the corporation's enterprise will be rewarded by greater profitability. The article hits on this association when it describes the "interdependence of a company and society" and "rather than merely acting on well-intentioned impulses or reacting to outside pressure, the organization can set an affirmative CSR agenda that produces maximum social benefit as well as gains for the business." (Porter and Kramer 2005)

Let us expand on Porter and Kramer's matrix of CSR "social linkages."

There is a hypothesis in developmental psychology referred to as the dynamic systems theory. At a basic level, the proposition suggests that human development is affected by an enormous range of environmental, social, genetic, and biological factors. In its radical state, personality and physical traits are not simply determined by the genes of one's parents and immediate surroundings as one is nurtured in early years, but that personality and physical traits are determined by the genes of great-great-great-grandparents hundreds of years ago and that most minute details of our surroundings and connections affect how we grow as healthy, responsible, caring, loving adults in society.

Porter and Kramer's linkages offer a CSR map of the corporation's role in society and how their interests, investments, and people contribute to the economic vigor and social well-being of the community. The corporation has its own dynamic systems that ensure growth and management and, in this case, integration with the developing community.

An effective CSR component of commerce, services, and industry has the most potential for effective change when the CSR effort is both strategic and systematic. Porter and Kramer offer a tremendous evaluation of strategy that incorporates interests in society, particularly addressing critical human issues and problems. This has a valid and perhaps crucial role in the corporation's internal profitability strategy.

The authors suggest that the purpose of CSR is not "moral obligation, sustainability, license to operate, and reputation." Nor is it really profit or public image at its heart. But, in fact, the purpose of programs in corporate social responsibility is to improve the human condition. The core values of entrepreneurship, commerce, services, and industry should include making the world a better place through the products a firm produces, the services it provides, and the value it adds to the global economy. Moreover, it should do so by allowing the firm to be actively and directly engaged with sustainable, systematic, and strategic development initiatives in the community, village, or neighborhood. By doing so in the most strategic and systematic manner possible, the corporation can assist society in addressing the most profound problems of the human condition such as hunger, poverty, and economic disparity. When we talk about strategy and systems, we are talking about society and, in its most salient and operable manifestation, the village – not just the corporation but the corporation's place in society and the global village.

Engineering change

President emeritus of the University of Michigan James Duderstadt's *Engineering for a Changing World: A Roadmap to the Future of Engineering Practice, Research, and Education* was written as part of the Millennium Project of the university in 2008 (Duderstadt 2008). The paper states that engineering and engineering education are both in a state of instability.

They are confronted by accreditation requirements, obstacles to greater diversity in the field, competition for grant funds, and insecurity about how to respond to increasing outsourcing of engineering talent and resources around the world. Duderstadt introduced the study by saying,

> Powerful forces, including demographics, globalization, and rapidly evolving technologies are driving profound changes in the role of engineering in society. The changing workforce and technology needs of a global knowledge economy are dramatically changing the nature of engineering practice, demanding far broader skills than simply the mastery of scientific and technological disciplines.
>
> (Duderstadt 2008: iii)

Duderstadt continued

> Bold transformative initiatives similar in character and scope to initiatives undertaken in response to earlier times of change and challenge ... will be necessary for the nation to maintain its leadership in technological innovation. The United States will have to reshape its engineering research, education, and practice to respond to challenges in global markets, national security, energy sustainability, and public health. The changes we envision are not only technological, but also cultural. They will affect the structure of organizations and relationships between institutional sectors of the country. This task cannot be accomplished by any one sector of society.
>
> (Duderstadt 2008: 95)

Duderstadt's assessment of engineering repeatedly refers to how the profession reflects the values of this country or supports this nation in juxtaposition to the rest of the world, undoubtedly due to the traumatic stresses placed on employment and the unsustainably of various industries in the United States because of global competition and outsourcing. Duderstadt's fears are well documented. No one can argue with the statistics that show how many jobs have been lost to telecommunication industries in India or manufacturing technologies in Asia. The shortfall of Duderstadt's premise is that present-day opportunities are discounted rather than emphasized. It could be much more advantageous to consolidate the resources of various corporate practices *with* global partners and to *integrate* technologies, biomedical advances, materials engineering, and computer science *along with* global partners and not against them.

Perhaps this will not be the case in twenty or thirty years, but right now there is an opportunity to expand technologies, engineering applications, and medical advancements in developed economies where markets are on the verge of resurgence and where engineering applications are critical to alleviating poverty, overcoming pandemic disease, providing secure shelter, and feeding 6,796,443,554 people *while* at the same time establishing a viable infrastructure for economies that will thrive on a global scale twenty years from now. It behooves educational institutions and industries alike to combine resources and invest in the infrastructure that will support the global base economy, especially and most importantly the economy that so poignantly affects the extreme poor, not just the United States economy.

The seeds of investment are in the minds of prospective young professionals and executives matriculating in our institutions of higher education. Investing in the infrastructure that will support the global economy in the future means involving industries and businesses with rising young professionals in educational and outreach programs in the villages of Kenya,

Vietnam, India, Malawi, Belize, and Haiti. Executives in our industries and the students they mentor bring innovation and enthusiasm to the field. They challenge the apprentice computer engineer to install a new network among computers in 174 schools in Belize to augment internet access and resources. They challenge the student civil engineer to design a water system to reach all 1,100 families in Bella Vista, Belize. They challenge the young mechanical engineer to build a regulator for steam heat systems in orphanages in Russia.

A corporate partnership with institutions of higher education and a plan to raise communities from the throes of poverty in a bold five-year program will secure not only America's future but the future of those emerging economies. Just as one village cannot overcome poverty alone, one executive cannot reverse the course of political and economic influences that perpetuate poverty. But one executive vested with the resources of a major corporation involved in a village partnership working with eager colleagues and perhaps an industry mentor will foster changes beyond her imagination for years to come – both in the village and in her career. Real world applications will promote genuine social change.

There are two ways of looking at conditions of poverty and disparity in our world. One way is to simply acknowledge that these conditions exist and to think that there is not much that can be done about it or that it is someone else's responsibility. The dynamics are too complex, the governmental and social obstacles too complicated, the global reach of poverty too overwhelming. Even if one could work on some of these problems, the results would be too modest to be effective.

On the other hand, effective programs in social responsibility look at conditions of poverty and disparity and envision how those conditions can be changed, which allies and resources can be marshaled in a uniquely strategic manner to address these conditions effectively and permanently, and who will provide the insights required to assess and analyze the existing dynamics that perpetuate poverty while also organizing potential resources, talents, and systems to alter those dynamics. The presupposition is that things can change and poverty can be overcome.

The convergence of resources and insights related to the alleviation of poverty is the catalyst of effective social change. In most cases, the insights that inspire and inform change already exist in the hearts and hopes of people who live in places where poverty is endemic. They are intimately familiar with the reasons that their neighbors and families are poor and they already have ideas about how to overcome those obstacles to achieve a realistic standard of living. In most cases, what they do not have are the resources to reach new standards and realize those development goals. They need that $200 micro-loan for a small business or to finish school.

The converse is true of people and institutions with wealth and resources. An institution, corporation, university, or even an individual may have far-reaching resources but lack the experience or insight to know how to effect change or create the best strategy. This is the basis of Paul Farmer's work, a world-renowned physician and head of Partners in Health (PIH). Farmer wants to introduce resource-rich institutions to communities where poverty is widespread.

What the authors are proposing is one step beyond the resources needed to rebuild schools, modify archaic heating systems, or distribute clean water to more families. The proposition is to build partnerships between those resource-rich and the places where people suffer, a partnership in which crucial human, networking, material, and financial resources are applied in novel, tactical ways with local citizens who alone have the experience and insight for social change.

Social change and economic transformation

Social change and economic transformation are inextricably bound together. It is not possible to have one without the other and both intrinsically benefit from the other as long as economic transformation narrows the divide between rich and poor.

Proctor & Gamble's former CEO Gurcharan Das explains how this happened in India over the past twenty-five years (Das 2006). Das explains that while there was tremendous economic growth in India, it was not restricted to the middle class but also raised a significant percentage of the poor above the poverty line every year in a prolonged democratization of resources. As Das points out, India's ascension has been based on domestic consumption and growth in services and technology rather than industrial production and commerce – two areas of the economy that actually present a much more formidable challenge to the rural poor for whom services and technology are not as easily utilized. Das calls for "a second green revolution" (Das 2006) in which India would invest more in agriculture, including renewed investment and developments in agribusiness and agritechnologies. While doing so, there has been significant growth not only in private enterprise (20 percent of Forbes 500 have R&D operations in India and 80 percent of all commercial loans go to entrepreneurs), but the private sector has also taken on typical public sector responsibilities such as education and infrastructure.

Herein lies the basis for CSR. Not only are there extensive social needs in India, that nation is a huge market and growing. The rate of domestic consumption of goods among India's 1.17 billion people is 50 percent higher per person than the People's Republic of China. Das estimates that at the current rate of growth, per capita income in India will catch up with the United States by 2066 (Das 2006). In the midst of this dramatic growth, there are enormous human needs and the slums of Mumbai and Kolkata rival the worst anywhere in the world. How can programs in corporate social responsibility respond to India's vast human needs and simultaneously pursue a corporate strategy in an economy dominated by service and technology? Seems to me like an epic opportunity for any corporation with sufficient resources.

Take the Loreto Schools, for instance, located in Kolkata, Darjeelikng, Delhi, Lucknow, and Asansol. This system of Catholic girls' schools is in some of the poorest areas of the country. They mix private education with services to the surrounding homeless. Modest corporate support would allow the Loreto School in Sealdah in Kolkata to enclose the roof area where children from the neighboring slums and railroad tracks are fed and educated. More significant corporate support would allow Loreto Schools to expand their integrated, value-based education to other underserved communities ("Where the Streets Have No Name" on www.youtube.com/watch?v=EOx1QTYLZQs). Easy to see what Das calls India's greatest needs – "providing basic education, health care, and drinking water" (Das 2006).

This reinforces the need for a system like the Peacework Village Network. In the Sealdah area of Kolkata or any other setting where poverty is endemic and where development needs are pervasive, it behooves corporate interests to ally themselves across disciplines. Contractors are most effective when they are working with others to improve infrastructure, provide educational resources from publishers, support technological improvements, implement new effective and fair legal and tax codes, develop strategies for improving nutrition among agribusiness entities, and collaborate with the non-profit and public service worlds. This systematic, sustainable, strategic approach to development and partnership will be, by far, the most successful and fruitful if the stages and overall plan are in response to locally identified needs with village leadership. Corporations should not manage the Loreto Schools, only provide supportive investment and infrastructure as Sachs suggested. It is the Loreto folks, not Proctor & Gamble or Nestlé, who know how to teach and how to reach the homeless children of Sealdah. This is the system by which the strategy of addressing social needs and profitability will work.

Developing a CSR approach

How should an organization approach entering the CSR domain? The answer is that a variety of factors contribute to the success of a CSR project. The critical success items include selecting the appropriate partners, in-depth pre-project planning, procurement, and project execution. The following section provides lessons learned in relation to each item.

Partners

Peacework has worked to improve schools, support literacy and education, and promote access to affordable and quality health care in Belize since 1994. Partnering with Peacework provided a network of individuals who helped to implement our ideas. This eliminated years of relationship building and enabled a rapid deployment allowing students to see the impact their efforts had upon the community.

Partners play the role of teacher, of constructor, of supervisor and of worker. They are invaluable to the experience, contributing to growth while instilling confidence in abilities.

Pre-project planning

A pre-project site visit is critical for identifying projects, building relationships with the community and identifying accommodations and material suppliers. A valuable lesson learned included the early involvement of the partners to provide feedback and identify constructability issues prior to arrival onsite. Structural stability issues were identified on the roof assembly for the water distribution storage huts in the field that could have been more easily addressed in the office.

Procurement

To ensure that the material required for construction to begin when the entire team arrives, a team member arrived three days early to coordinate with the Peacework staff onsite. This individual ensured that the correct material was procured and transported to the site. Material delivery time is often a large unknown in international settings. For short duration projects, it is best to have all materials delivered prior to or at the beginning of the work. However, if there is an assurance that materials can be delivered in a short period (next day or two), it may be beneficial to begin with a lower than estimated quantity. Additional materials can be procured mid-week if necessary. This will allow the group to absorb scope changes or overestimates.

Project execution

In order to execute the project, proper equipment is essential as well as site planning. Locating the material storage sites for large items such as block, sand and aggregate, is critical so they do not conflict with project locations. Equally important is the identification of an on-site storage location which affords security and allows overnight storage. In addition to material issues, it is important to consider the well-being of the project team. Many volunteers become exhausted mid-week. To ensure the safety of all individuals on the site, coordinate work schedules around normal office times and around weather (hottest times of the day) as much as possible.

The implementation of this program, while successful, also brought to light several lessons learned. These include selecting appropriate partners, planning the project, procuring materials and executing the project safely and efficiently. Based upon the success of this project and the lessons learned, the authors feel that with customization that this program can be successfully implemented in any corporate setting.

Conclusion

Programs in corporate social responsibility should seek allies with universities who may already be affiliated with their industrial, agricultural, or business services. Universities serve as a source of valuable experts, volunteers, and even potential future personnel. Institutions such as land-grant universities have a multidisciplinary mandate for regional and global economic development. J.W. Fulbright's charge to the University of Arkansas was to "serve Arkansas and the world." Prior corporate and university affiliations make the academic connection propitious.

Programs in corporate social responsibility will be most effective when they develop multidisciplinary initiatives, taking full advantage of alliances in commerce as well as fields from agriculture to law.

Programs in corporate social responsibility should be part of the overall business strategy of the firm, utilizing resources among varied personnel and the company's charitable trust or foundation as an integral part of the CSR plan. Why have one corporate division supporting health care in poor rural villages without coordinating that outreach with the additional finances that the company has preserved in its foundation or charitable trust? Should not the corporate trust and CSR be integrated in a more effective and efficient use of those collective resources?

Programs in corporate social responsibility should utilize experts in the field, perhaps an NGO partner such as the Peacework Development Fund with experience in community development and the global social, economic, and political maze.

Programs in corporate social responsibility recognize the extensive contributions that individuals in the firm are already making to society through civic clubs, service in community settings, on boards of local charities, or service through churches and synagogues.

Programs in corporate social responsibility serve as part of a larger campaign to alleviate conditions of extreme poverty in our world, to devote at least a portion of their resources to aid folks at the very, very bottom. This involves coordinating efforts with respected independent campaigns such as the UN's Millennium Development Program, collaboration with the World Business Council for Sustainable Development, participating in the World Economic Forum, or supporting the Academy of Management's Alleviating Poverty Task Force.

Programs in corporate social responsibility foster positive changes in global economic policy and international cooperation. The International Monetary Fund (IMF) is well known for advancing a development policy called Structural Adjustment. The long and the short of it is that structural adjustment means poor villages ought to adjust their economic priorities and development objectives to match First World policy expectations and debt demands when, more appropriately, the reverse would be more effective and democratic. First World policies and financing ought to match economic needs and development objectives tailored to each local situation. If privatization is one of the most pressing objectives of Structural Adjustment, who better to institute those changes than private enterprise in support of genuine social change and effective economic development? They key is bottom-up economic development where people are intimately familiar with their own conditions. It is my theory that bottom-up development using strategic and appropriate resources will lead to positive, sustained, systematic social change.

Programs in corporate social responsibility are part of the core operation of the business acknowledging the inherent, albeit in many cases long-term, strategy and profitability of sustained, solid, responsible corporate citizenship.

Programs in corporate social responsibility support the core of the business and never lose sight of the fact that social responsibility ultimately is improving the human condition.

References

Aaker, J. (1993), *Partners with the Poor*, Friendship Press, New York.

Das, G. (2006), The India Model, *Foreign Affairs*, July/Aug, pp. 2–16, online, available at: www.foreignaffairs.com/articles/61728/gurcharan-das/the-india-model.

Duderstadt, J.J. (2008), *Engineering for a Changing World: A Roadmap to the Future of Engineering Practice, Research, and Education*, online, available at: http://milproj.dc. umich.edu/publications/EngFlex_report/index.html.

Porter, M.E. and Kramer, M.R. (2005), Strategy & Society, The Link Between Competitive Advantage and Corporate Social Responsibility, *Harvard Business Review*, 84(12): 78–92.

Sachs, J. (2005), *The End of Poverty: Economic Possibilities for Our Time*, Penguin Press, New York.

12 Conclusion

Putting the issues in perspective

Paul S. Chinowsky and Anthony D. Songer

No organization can maintain "status quo" indefinitely and remain competitive in today's changing business environment. The need for improvement and change are varied and heavily dependent on organization culture, market demands, competition, workforce demographics, long-held industry practices, and a necessary focus on minimizing risk. Even with these varied forces working for and against change, an organization must decide to improve and modify procedures on a regular and continuing basis. The need for this improvement in response to changing environments is a recurring theme through business literature in particular and all areas of production in general. As outlined by Collins (2001), the greatest threat to achieving greatness is an organization that is content with being very good at what it does. In the context of strategic issues, this thought transfers to the greatest threat to greatness being contentment with established practices that have achieved acceptable and above average results. This focus on moving from a position of being successful in a specific area or context to one that potentially excels above the competition can be daunting to many organizations. The cost-benefit analysis for introducing a disruption to existing processes can be difficult to establish due to the unknowns involved with the process. However, this is where leadership is most required. Leadership is required to pull the organization in the same direction, set the vision, and synthesize the varied internal and external forces that are influencing the organization.

With that said, we do not want to end this book with the familiar refrain that organization success is solely dependent on the effectiveness of senior leadership. Is this a major factor? Absolutely, but it is not in any way the only factor. Rather, transforming an organization into one that achieves goals that exceed traditional expectations requires a combination of focal points. Understanding the economic environment in which the organization operates is just as critical as putting in place a good leadership team. Similarly, building a strong network in the organization that is willing to share knowledge is a fundamental element of a strong organization. Once this network is in place, the organization can examine the global environment, the competitive situation, and the emerging markets that present the opportunity for growth and continued success. Each of these elements is interdependent on each other in terms of building sustainable success. Although leadership is required to guide the process, every member of the organization is responsible for providing an analytical, strategic, or operational input to the overall success picture.

The next steps

Now that we have eliminated the common ending point of placing the burden of success solely on the shoulders of the organization leader, the responsibility is now there to answer the issue of what are then the next steps for the organization. In response, we conclude the book by emphasizing one last topic, implementation. Any change in the organization is going to require implementing a new initiative, practice, or set of strategic goals. However, this implementation effort is often one of the most difficult and often serves as the barrier to achieving strategic success. The core of this challenge is often found to be in the preparation component of the process. Specifically, organizations do not spend the time required to ensure that they are prepared to embark on a new strategic goal or process (Chinowsky 2008). This lack of focus results in the lack of a clear perspective on where the new initiatives are going to lead and a lack of focus on the overall steps required for the entire implementation process.

Therefore, to provide a conclusion to the overall perspective on total organization improvement and leadership, we present eight key areas that organizations should consider when embarking on an improvement process in any of the eight areas presented in this book. These eight areas can be considered as the initial implementation roadmap where each area needs to be a destination that is visited and completed before the implementation journey can successfully commence. The eight areas are as follows:

1. **Vision** – A vision provides the organization with a goal for the completion of the implementation process. A successful strategic implementation requires the organization leaders to create a clear, short vision for the implementation process.
2. **Support** – A successful change in organization direction requires support from all levels of the organization, from top leadership through professional staff. This vertical support throughout the organization must be put in place prior to the commencement of the change process.
3. **Communication** – Communication is the foundation for any new strategic effort. Unfortunately, this is one area which organizations often fail to completely address. Specifically, a failure to keep organization personnel informed throughout the process will result in misunderstandings regarding the strategic initiative and may result in a failure of the process.
4. **Roadmap** – A roadmap describing the process and the milestones during the implementation of a new direction is necessary to keep all personnel moving in the same direction at the same rate. The lack of this road map will lead to confusion and miscommunication as people move in different directions and aim for different goals.
5. **Necessity** – Why introduce a new practice or strategic goal if the existing operations work the way they are intended? This is the common response to new strategic initiatives. Therefore, preparing the organization for a new initiative by establishing the necessity for the initiative is a foundational element in the implementation process.
6. **Champion** – Every initiative requires an individual(s) to lead the implementation process. This leader will push the strategic effort and subsequently also receive the greatest resistance to the change effort. *Thus, the individual(s) selected to fill this role will need to have the communication skills to overcome the resistance.*
7. **Empowerment** – The empowerment element focuses on the need for individuals to have the responsibility and authority necessary to make incremental changes to support the strategic initiative. The question for the organization is whether individuals at all levels have the authority to make the incremental changes required for overall success.
8. **Education** – The organization must have sufficient knowledge of the proposed initiative and the impact of the change to overcome resistance to the practice. This knowledge can only come from education opportunities provided to employees at all levels.

The attention to these eight elements will provide a significant boost to an organization's ability to introduce a new strategic objective. However, preparation only gets you to the starting line. Any effort will require the successful completion of the journey to confidently move forward with a strategic effort. Although many authors have provided guidance to organization leaders on how to complete this journey, and this topic is outside the scope of this volume, a few key elements are worth noting as an organization prepares to embark on a strategic effort. Specifically, the following elements should be considered.

1. **Needs analysis** – An organization should strongly consider conducting a "needs analysis" that lays out the case for implementing a new initiative. This case will need to be presented to both management and critical staff, so a compelling argument is required as an output of this step.
2. **Develop plan** – Once a needs analysis is completed, the organization should explicitly develop the details of how the strategic initiative is going to be achieved. These details should then be translated into a plan that meets the requirements of the specific organization.
3. **Develop milestones/tasks** – A plan is only as good as its ability to guide the organization in terms of milestones that need to be reached. Therefore, a set of achievable milestones and tasks that have reasonable goals and dates is essential to completing the plan. Try to avoid setting more than three tasks per milestone or the organization could be setting itself up for disappointment or even failure.
4. **Implement tasks/plan** – Completing tasks and meeting milestones is an iterative process. This is the phase where oversight is critical in the process and thus the organization should focus on assigning champions and supervisors to critical tasks. The milestones should not be left to be completed by themselves.
5. **Perform evaluations** – In conjunction with Step 4, Step 5 requires the team to perform evaluations during each phase of the process to determine if the process is achieving the anticipated goals. Do not be afraid to re-evaluate and redirect the process. This is the future of the organization at stake and getting it right is more important than admitting that an adjustment needs to be made.
6. **Benchmark** – The final step in the process is to benchmark the new initiative both internally and externally. The intent of this benchmark process is to determine if the initiative is returning the results that management intended to obtain. The specific benchmarks that are used in the process will be determined by the individual organization. However, a big caveat here. Do not fall into the trap of benchmarking against organizations that are achieving at a mediocre state, but which are considered advanced in your sector. This is a trap that often hinders organizations where they look around at competitors and say we are doing just as well as they are. Unfortunately, none of the competitors are performing at a level that you would like to achieve. This leads to a false sense of security where you have just set the bar lower than intended. Do not be afraid to say that you want to achieve more than the competition. This is your organization – set the goals where you want them to be!

Final thoughts

The readiness areas have been put on the schedule, the roadmap with milestones has been created, the key strategic initiatives have been selected; it is time to get started. Or is it? Have we forgotten anything? Yes, one key item. The people. We know we said that we were not going to leave this volume with the impression that everything is dependent on the leadership and we are going to remain committed to that. However, we are going to leave you with a final thought that *is* people related. Specifically, to remember that the success of any strategic initiative in a service-oriented business such as engineering and construction ultimately comes down to the people who are implementing the initiative. This is not limited to leadership, but is inclusive of all members of the organization. A successful strategic initiative requires all members of the organization to be engaged. More than just engaged. It requires every member of the organization to believe that long-term success is dependent on each person being an active contributor to the process. True, this does require leadership to put this in place, but ultimately, this is going to be dependent on the organization as a whole wanting to succeed. That is the thought we want to leave you with. Success is not dependent on one initiative or one person. You cannot take one chapter in this book and say that is the key to success. The issues presented here are part of a whole that needs to be perceived as a comprehensive approach to long-term success. The combination of this perspective with a belief that every member of the organization needs to be committed is in fact the first step to achieving sustainable success in an industry which is undergoing significant and permanent change on a global stage.

References

Chinowsky, P.S. (2008), "A staircase model for successful practice implementation," *Journal of Management in Engineering*, ASCE, 24(3): 187–195.
Collins, J. (2001), *Good to Great*, HarperCollins, New York.

Index